Learning Native Wisdom

Culture of the Land
A Series in the New Agrarianism

This series is devoted to the exploration and articulation of a new agrarianism that considers the health of habitats and human communities together. It demonstrates how agrarian insights and responsibilities can be worked out in diverse fields of learning and living: history, science, art, politics, economics, literature, philosophy, religion, urban planning, education, and public policy. Agrarianism is a comprehensive worldview that appreciates the intimate and practical connections that exist between humans and the earth. It stands as our most promising alternative to the unsustainable and destructive ways of current global, industrial, and consumer culture.

Learning Native Wisdom

• • •

What Traditional Cultures Teach Us about Subsistence, Sustainability, and Spirituality

GARY HOLTHAUS

THE UNIVERSITY PRESS OF KENTUCKY

The University Press of Kentucky
Scholarly publisher for the Commonwealth, serving Bellarmine University, Berea College, Centre College of Kentucky, Eastern Kentucky University, The Filson Historical Society, Georgetown College, Kentucky Historical Society, Kentucky State University, Morehead State University, Murray State University, Northern Kentucky University, Transylvania University, University of Kentucky, University of Louisville, and Western Kentucky University.
All rights reserved.

Editorial and Sales Offices: The University Press of Kentucky
663 South Limestone Street, Lexington, Kentucky 40508-4008
www.kentuckypress.com

Library of Congress Cataloging-in-Publication Data
Holthaus, Gary H., 1932–
    Learning native wisdom : what traditional cultures teach us about subsistence, sustainability, and spirituality / Gary Holthaus.
        p.   cm. — (Culture of the land)
    Includes bibliographical references and index.
    ISBN 978-0-8131-2487-2 (hardcover : alk. paper)
    1. Environmental ethics. 2. Sustainable living. 3. Indians of North America. I. Title. II. Series.
    GE42.H65 2008
    179'.1—dc22
                                2007052577

Manufactured in the United States of America.

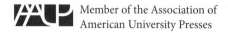

Member of the Association of
American University Presses

For Robert D. Arnold
(1931–2003)

Dorik V. Mechau

Ted Chamberlin

Gary Snyder

Colleagues, friends, inspirators, conspirators, mentors, participants in
many adventures, longtime companions along the way

Contents

Axe Handles

One afternoon the last week of April
Showing Kai how to throw a hatchet
One-half turn and it sticks in a stump.
He recalls the hatchet-head
Without a handle, in the shop
And go gets it, and wants it for his own.
A broken-off axe handle behind the door
Is long enough for a hatchet,
We cut it to length and take it
With the hatchet head
And working hatchet, to the wood block.
There I begin to shape the old handle
With the hatchet, and the phrase
First learned from Ezra Pound
Rings in my ears!
"When making an axe handle
        the pattern is not far off."
And I say this to Kai
"Look: We'll shape the handle
By checking the handle
Of the axe we cut with—"
And he sees. And I hear it again:
It's in Lu Ji's *Wên Fu*, fourth century
A.D. "Essay on Literature"—in the

Preface: "In making the handle
Of an axe
By cutting wood with an axe
The model is indeed near at hand."
My teacher Shih-hsiang Chen
Translated that and taught it years ago
And I see: Pound was an axe,
Chen was an axe, I am an axe
And my son a handle, soon
To be shaping again, model
And tool, craft of culture,
How we go on.

—*Gary Snyder*

Introduction

Why Subsistence, Sustainability, and Spirituality?

WHY LEARN NATIVE WISDOM? Because if we want to think about a sustainable culture and find ways to create one, we have models right at hand, as Gary Snyder indicates in "Axe Handles." The models have roots and forms in several cultures, but in each case they grow from antiquity. I'm thinking here of Eskimo and Indian cultures in Alaska, whose roots are older than even those Chinese sage kings who preceded Confucius, and may be three times as old as Western culture. They have been around long enough; there must be something we can learn from them, if we have sufficient humility and wisdom of our own.

What are the links among subsistence, sustainability, and spirituality? We are in a fix, caught in a species die-out at a rate never before seen. Our earth, upon which we remain totally dependent, has become infected by its own chronic wasting disease, losing its "ten thousand things," as the ancient Chinese called the myriad creatures, plants, animals, and rocks that cover the planet; we have gotten caught in our own desires, greed, conceits, and arrogance. And there lies the heart of it. The evidence seems clear now that the real root of these issues, both cause and cure, lies not in our science or technology but in our own spiritual and intellectual poverty—or, more hopefully, in our own spiritual and intellectual resources. Those Neo-Confucians from around the twelfth century would have called these "the heart-mind."[1] Mary Evelyn Tucker, one of the great scholars of Confucianism, sums it up in a sentence: "Nature is seen as a 'resource' to be used rather than a 'source' of all life to be respected." She concludes that our division of life into matter and spirit, and our sense that spirit is superior to matter, "has given rise to a crisis of culture, a crisis of the environment, and a crisis of the spirit."[2]

There is a paradox at work here that puzzles many of us. Religions are booming around the globe, with adherents willing to lay down their own lives, and the lives of many others, to foster faiths that are not alive and awake to their own best spirituality but are now given to revenge, power, violence, and excess. All the great religions and philosophies and their indigenous forebears that we know of insist that revenge, power, violence, and excess prohibit our living out a faith that must include others in the arena of our concern—must protect the weak, feed the poor, and treat the ten thousand things with deepest respect. We are failing on all counts while politically and socially powerful persons who practice Christianity, Islam, and Judaism ignore their own most basic precepts and subvert their religions' best intentions and hopes. We are fortunate that there are cultures far older than those of the desert trinity from which we can learn another wisdom that may save us yet.

This book is not a how-to guide that shows us some easy way to get out of this mess. It contains essays, meditations, about the concept of sustainability. It seems important to me to consider the concept because my own experience in small towns in the Midwest and West leads me to think the concept is not widely understood. If we do not understand sustainability, we cannot live toward it. If we do not live toward it, we have a foreshortened future that leaves our children in jeopardy. Yet, as I listen to friends in business, I find that many still see sustainability as a radical environmentalism that threatens commerce. They seem to believe that economics is the great determiner of every human action and that a growing economy is our only avenue to a viable culture. That view ultimately destroys the land, impoverishes the largest number of people, and reduces our spirituality to deciding which religion pays the best, makes us most comfortable with our wealth, or helps us gain affluence and influence friends. On the other hand, friends working to protect the environment or create sustainable communities or sustainable agriculture often see the local chamber of commerce, the powerful elites who now rule the country, and especially transnational corporations as impediments to, if not enemies of, a sustainable future. Folks on each side of that divide work hard to thwart and punish their colleagues on the other side. I believe those practices work against our hopes that we might have a sustainable future and that our children might have decent lives.

I have put sustainability between subsistence and spirituality in the subtitle of this book because, although all three are equally urgent, sustainability is the central issue I am trying to understand. All cultures are subsistence cultures, though we Western descendents of the Enlightenment no longer recognize ourselves as a subsistence people utterly dependent on the land in the same way that a Yup'ik Eskimo, for example, is. This seems to fly in the face of anthropology, which reduces indigenous peoples' life on the land to an economy. Some speak of subsistence peoples living at a "*mere* subsistence level," which to them is mean or even deprived. That has not been my experience of people living that life. That subsistence life we recognize among traditional cultures risks occasional hunger, no doubt, but it runs no more risk of hunger than our own system, which has produced ample food for all and yet left millions starving, and millions more—even in our own nation—without food security. Village people feed themselves pretty well and far more equitably by comparison. Some think our cash economy distinguishes us from more "primitive" cultures. Such talk clearly means that current American culture is not primitive, is not a subsistence culture. Yet talking with farmers, as I did for *From the Farm to the Table: What All Americans Need to Know about Agriculture*, one thing becomes abundantly clear: we are still as dependent on soil health, on our land and waters and what they produce, as any culture that ever existed.[3] Thus all cultures are subsistence cultures; not all cultures are sustainable. We can turn our backs on our subsistence life and deny it, as our culture has tried to do. But the result of that denial will always be an ignorant and exploitative use of the soil and other resources, no matter how advanced our technologies may become. We can never become a sustainable culture if we deny our own dependence on subsistence, our dependence on the land. Our technology will not save us, and it is not separable from the earth, despite our faith in it. Our soil is closer to our salvation, if we can overcome our tendencies to ignore and abuse it.

The most important task in our time is not to protect the land or create social justice but to create a sustainable culture. Why try to create what Wisconsin farmer, teacher, and part-time philosopher Prescott Bergh calls a sustainable culture? Prescott first used that phrase, in my hearing at least, when a small group of farmers and businesspeople got together to talk about organizing a conference on sustainable agriculture. At the beginning of that meeting, Prescott said, "There's not much point in talking about sustainable agriculture if you don't have a sustainable culture to back it up, and America doesn't." We are so consumer oriented, he said, that we are destroying ourselves, eating ourselves out of house and home. His statement revealed to me how narrow my own vision of sustainability was. I had to look at sustainability on a much broader and deeper scale than I had conceived. One thing is clear: the path to a tolerable future does not lie in continual economic growth. That is a fatal path to follow. A path to a tolerable future inevitably leads through subsistence and spirituality and works toward sustainability. This book tries to shed some light on that path by speaking to the concept rather than providing another how-to look at what needs to be done. It is, appropriately I think, about relationships, in this case the relationship between spirituality and subsistence, and the relationship of each to sustainability.

I believe that our spirituality is at the core of the problems we face, whether they are environmental or in the realms of economics, social justice, and simple humaneness. This is not a new idea. But I have yet to attend a meeting, lecture, or workshop on sustainability that addresses it. It gets mentioned occasionally, and there are murmurs of assent, and then we quickly turn back to the more "practical" and more comfortable work that needs to be done. One sign that spirituality is the most fundamental issue is that we prefer to work on changing others, through war if necessary, rather than changing ourselves. Both the Qur'an and the Bible have something to say about that: "Allah does not change a people's lot unless they change what is in their hearts, " says the Qur'an (13:11). "Don't look for the mote in your neighbor's eye while you have a beam in your own," warns Jesus (Matt. 7:5 [Revised Standard Version]). Confucius, too, considered this. Working to transform oneself toward a wise and sagacious character, which Confucius called "self-cultivation," is a primary task in the Confucian traditions. Until we have ourselves in order, Confucius holds, our families, our community, our state, the very stars in their courses will be in disarray.[4]

Questions about subsistence, sustainability, and spirituality will come up again and again in various forms in what follows. This is in part because of my own uncertainties. I have to wrestle with these matters myself, and I assume others do as well. The issues caught in Prescott's statement are so huge, so complex, that for a time I hardly knew where to begin. My thought for now goes like this: Where to start, obviously, is with myself, not the whole world. If I can get myself right, if we can get ourselves right, as our most reliable forebears from many cultures have shown us, the culture may begin to tilt toward a more sustainable character. Thus spirituality, my own in this case, your own in your case, is one key to creating a sustainable culture. Reconceiving the subsistence life is a second step. Out of a combination of those efforts, we may be able to take a third step: to work toward becoming a sustainable culture. If we have time.

These essays are my attempt to look squarely at subsistence, sustainability, and spirituality. Experience in a variety of jobs and an even greater diversity of volunteer activities has led me to think that my concerns for the environment and social justice are linked and that the key link is a spiritual and intellectual one. Though taking care of the environment and creating social justice are not the most important tasks, they are absolutely essential components of a sustainable culture. With a clear understanding of our spirituality and a clear view of the subsistence nature of our society, sustainability takes on a form that we can all understand and work to create.

I begin with this confession: I believe that our spiritual lives are rooted not in creeds and scriptures or particular beliefs or rituals but in our use of language and stories. Language shapes thought and behavior and informs our spiritual life. We create balance and harmony, Confucius says (as I read him), by clarifying our words. Both government and persons function best by "the rectification of names." Epictetus, the first-century Greek philosopher, echoes Confucius and his own Stoic colleagues when he says, "One of the clearest marks of the moral life is right speech. Perfecting our speech is one of the keystones of an authentic spiritual program."[5] He does not need to add that those who deliberately misuse words cannot provide moral leadership. In the Christian tradition, John's gospel (1:1–3) tells us that everything is created by the logos, the word, a notion that was abroad in many cultures long centuries before John.[6]

These essays contend that the issues we face are issues we have created for ourselves and that with self-discipline, perseverance, humility, and considerable luck we may yet rectify them. And that we will never create a just and justifiable economy or a just society, or treat the environment rightly—all essentials for a sustainable culture—until we find a spiritual life that fosters greater respect for our neighbors, including our neighbor trout, as poet Richard Hugo puts it, our neighbor tree, our neighbor bird.[7] The key to that spiritual life finally lies in getting our words straight and using them to create healthy stories, songs, and poems. Clearly, one thing I am trying to do here is organize in a coherent pattern the thousand ideas that have moved through my head from myriad sources over the years. The question I've had to wrestle with is, How do these notions relate to one another to provide a way to a sustainable culture?

Though the spiritual element of a sustainable culture remains impractical in the common view, there is an urgency about it. Our primary concern at the moment, the loss of fossil fuel, is but one of the forces converging on us now. Climate change is another and perhaps far more devastating concern. Loss of irrigation water, groundwater, and drinking water is another. Soil loss, and the loss of nutrition in the soil that is left, brings additional perils. The decline in fossil fuels will inevitably be marked by an accompanying decline in any food supplies that are dependent on long-distance hauls and oil-based pesticides, herbicides, and fertilizers. With the loss of oil come the loss of plastics, the loss of computers, and the loss of electricity. Our racism, ethnocentrism, sexism, and homophobia, and the ever increasing gap between the rich and the poor, bode to be as essential and as difficult to resolve. Each of these is but a cultural fragment; none, destructive as they may appear, are fundamental. Each represents only part of our task, and we must look at the whole as well as these fragments. Alas, America is no longer a moderating force in these matters but an exemplar of their persistent virulence and a trigger for increasing them. As these forces converge on us, violence is apt to accompany them. A 2003 report prepared for the Department of Defense warns that climate change will be accompanied by escalating political violence and social discord. The authors conclude, "Disruption and conflict will be endemic features of life."[8] It is most practical then to work on the most fundamental elements of our humanity while we work at the most pragmatic tasks before us: not to create a strong economy but to figure out how to survive and feed ourselves.

We may not be able to turn aside the troubles facing us, but we can cultivate a spiritual life, creating an individual and social spirituality that may short-circuit the violence and lead us to take care of one another and the earth. If we do not work on that fundamental, the worst is sure to come. But history also shows us that difficulty can unite us in a common cause, one aimed at creating a sustainable culture, avoiding the violence, and not just surviving but emerging from the conjunction of these forces more fully human, more respectful of one another and the world, than we have ever been in the long history we can trace. I hope it is clear that developing the spirituality I propose does not require giving up any religious belief or personal faith. You can comfortably remain a Muslim, Christian, Buddhist, Zoroastrian, Hindu, animist, or whatever is most personal and still adopt a spirituality that supports sustainability. I am urging a spirituality that leads to sustainability; my outline of it may not be yours. Between us, respecting each others' efforts, we can keep working on it. If we do, we may yet have a future as a species and as a global family.

I call these "vernacular essays" because their tone is oral; they are as close to my own voice as I can make them. They are also vernacular because they are inclined toward the anecdotal, for I believe in the power of stories as a principal means to transform us as individuals and as a culture. I know myself to be embedded in the mainstream American culture, but "mainstream American culture" is a phrase I don't often use in conversation because it is an abstraction that distances us from our participation in the culture. Often in these essays I refer to "our culture" or "our American culture," which may lead some readers who feel alienated from the mainstream to think this book is not for them. They may say, "Well, it may be his culture, but it sure isn't mine." Yet I persist because even the most alienated among us are also participants in this culture. For those of us who live here, it cannot be avoided.

I remember, years back, early morning conversations with homeless Eskimo men in Anchorage. They might seem to be nonparticipants, cultural driftwood abandoned by the mainstream, stripped of all their bark: no driver's license, no bank account, no voter registration, no home, and even on some cold days no jacket. Not always sure of their next meal but knowing all the Dumpsters in the alleys behind the fine restaurants and hotels on Fourth Avenue. Yet I listened to them talk about how it was when they were in the army in World War II or the Korean War. *Enlisted* men. Participants. One told about being kicked off the bus near his base in the South because "the driver thought I was black. He thought I was black. No, I tell him, 'I'm Eskimo. I'm Eskimo.' He don't know Eskimo. He made me get off the bus. He never heard of Eskimo." Everyone bent back with laughter. Once another said, perhaps remembering Korea, "You get shot, you feel nothing. I got shot, didn't feel nothing, just spin around and fall down." Another took it up, "Shock. That's shock. You don't feel nothing, that's right. Just fall down."

Wounded expat participants in our culture, these men camped under Visquine in a gully above the creek down near the "native" hospital. They showed me how they kept their sleeping bags dry. Twine between the trees, the plastic draped just so, a couple of cement blocks from who knows where holding down the edges. "Works pretty good," they told me. Our conversations often took place over Egg McMuffins in the Mc-Donald's on the corner of Fourth and C at a little after six o'clock in the morning, when the place opened. They came in to warm up, hunched over, hugging their mugs of coffee with both hands, bowing their heads over the cups, breathing in the vapor. Participants still. What could have been more American in 1980 than hanging out in McDonald's?

I have been in court on Monday mornings when they were under arrest, again, for holing up in an abandoned building during a cold spell, or for being drunk. The judge read those still dazed men sitting in the docket, chins on their chests, their rights in English, a second language for all, and that at a third-grade level. The judge spoke so quickly to get the ritualized announcements over that I could not understand what he said though I was sober: "I'm-going-to-tell-you-your-rights-and-I-want-you-to-listen-carefully-because-I-am-going-to-tell-you-one-time-and-one-time-only . . ."

Those men may never read this book or live the way I do, yet they are participants in our culture. Regardless of how we want to disown it, at some level we all participate, and it is ours. So I write of "our culture," meaning all of us who live here. I want everyone to feel ownership in the things I describe. I want to be up front and personal. It is not *the* government that is creating the "consume, consume" din in our ears but *our* government. I describe some unhealthy stories that prevent sustainability, but these are not stories that mainstream Americans have been telling themselves. They are stories, alas, that we, all of us, have been telling ourselves. It was not abstract Western institutions that practiced ethnocide in our public schools in villages across Alaska. It was me, teaching in Naknek, imposing myself and my culture on students whom I knew personally and whose elders spoke a language that was not mine. I don't want any reader thinking it was mainstream American culture doing it; it was me, my culture, my fellow teachers, our culture, *we* who were doing it. At Sand Creek, it was not *the* militia but *our* militia that returned with genitalia stretched over their saddle horns, and it was we who applauded and cheered Colonel Chivington and his men. That's us too. Not our best us, but us. Even when our government does things we would stop if we could, it's our government. Our government, our culture does things every day that I despise. But the fact that I despise those things does not mean that the government is not mine or that I am not a participant in the culture.

The good thing is that we also know there is a much better "we" available to us every moment. We see that finer participation in our culture every day in friends who suffer loss but are never diminished, in coworkers who have a compelling vision of how the world might become a better place and work toward it for all they are worth. We know people who have found their place and are at home, even in this world, with all its confusion and devastation, and work to clarify our vision and end our pain. We know people who, without any deliberate effort, help us to become better than we would have been if they were not in our lives. We see it in ourselves when we find worthwhile work or act on our own generous impulses. You will find some gleaming figures of hope in these pages too. They also are us, and they, too, represent our culture at work.

"Our culture" in my usage includes Christianity, Islam, Judaism, Shintoism, Buddhism . . . They are all us too, a common heritage. "Our culture" includes all ethnic members, folks of all sexual orientations, visitors on visas who are but temporarily "ours," immigrants from other nations. It's all America, and each of us is caught up in the same social and environmental processes.

Only if we acknowledge our participation can we protest against whatever is unhealthy, unwise, unfair, unhallowed, unsustainable. We are all in this together, including my Eskimo friends telling their stories of participation and alienation in McDonald's and living under transparent plastic tarps in the gully. They were fine fellow participants in our culture.

So these are mostly stories, vernacular stories that come from my own experience, from my conversations with friends, and from my reading. In that sense they are genuine essays, a means of thinking my way into these issues. As essays they are exploratory and tentative. They are an invitation to conversation as well as a means for me to find out what I really think. At least for now.

Back to Basics

Music and Story

WHY TRY TO CREATE WHAT Prescott Bergh, the Wisconsin farmer, calls a sustainable culture? Because the disparate efforts that we are making now, though essential, are too small for the task that confronts the world. Our myriad efforts to create a sustainable aspect of our lives together include sustainable agriculture, sustainable communities, sustainable energy, sustainable economies, sustainable ecosystems, sustainable bioregions, sustained yield in natural resources like timber or fisheries . . .

But apparently no one is trying to bring all those efforts together, to think about ways that we might create a sustainable culture that will take into account not only humans but also all other creatures, the health of plants and soil and landforms. Richard Norgaard, professor of economics and environment at the University of California, Berkeley, author of *Development Betrayed*, and chair of a United Nations nongovernmental organization special committee on sustainability, remarked to me in 1998 that he had hoped his committee would fulfill that larger integrative function but that he had been disappointed in that dream. "Everyone is off, busy at their own tasks," he said; "no one is keeping an eye on the big picture."

Further, the folks working on sustainable economies do not often talk—at least at any length or depth—with those who are working on sustainable communities, doing sustainable agriculture, developing sustainable yields in timber or fish, monitoring air or water quality, or creating curricula for our schools that will help our children take better care of the world than we have. Not many focus on the task that transcends those, on how we might create a sustainable culture, a system that will incorporate all those efforts and restore a semblance of wholeness to our cultural life. And few talk about the spiritual nature of such an enterprise, the issue that lies behind our efforts to create environmental health, social justice, and a sound economy. How ironic it would be if those whose first principle is that everything is connected were to fail to create a sustainable whole because we failed to connect with our colleagues working toward the same end.

We can create a sustainable agriculture and still have an unsustainable planet. We can create thousands of sustainable communities and be left with an unsustainable world. We can build sustainable economies and still lack other essentials that will create a sustainable life for all. And who, except the self-destructive, wants to work toward an unsustainable culture?

Since Prescott posed his question about a sustainable culture, others have reinforced it: "Thus the agricultural assumption that nature is to be either subdued or ignored is embedded in the larger cultural assumption. Therefore we should not expect sustainable agriculture to exist safely as a satellite in orbit around an extractive economy," writes Wes Jackson in *Becoming Native to This Place*.[1] Creating a sustainable agriculture, a sustainable economy, or a sustainable community or making any other sustainable effort is not an end in itself. Each is a means to that larger end of creating a sustainable culture. If a sustainable culture is the goal, as Prescott said, or a regenerative culture is our best destiny, as Ben Webb, an environmentally minded Episcopal priest in Iowa, suggested, it might be useful to take a look at really long-lived cultures to see just what characteristics might have contributed to their longevity. Perhaps other cultures can teach us what we cannot seem to learn for, or from, ourselves.

Wes raises other interesting questions in *Becoming Native to This Place*. Writing about his home county in central Kansas, he points out that, according to archaeological evidence and observations recorded in early explorers' journals, in the sixteenth century more than 25,000 people lived within what is now Rice County. The land then, Wes says, supported about 35 persons per square mile. The first settlers came with their plows in the mid-nineteenth century, and by 1927 the population was reduced to 15,000 people and falling.[2] By 1935 the topsoil was blowing away, and dust was clogging the buildings, machinery, and everyone's lungs. The great photographer Margaret Bourke-White wrote of watching cattle "run around in circles until they fall and breathe so much dust that they die." It got so bad, she wrote in May 1935, that "farmers dread the birth of calves during a storm. The newborn animals will be dead within twenty-four hours."[3] Such farming practices as Americans had implemented by the end of the nineteenth century were not sustainable, and we did not make them sustainable in the twentieth. According to Wes Jackson, the Rice County, Kansas, population fell to just 10,400 in 1990. "Why this huge decline in numbers of people? Were the natives more sophisticated at providing their living than we are?"[4] Good questions.

I am not a trained anthropologist, though I am familiar with the literature of that field, but my experiences in Montana and Alaska lead me to suggest that we have something to learn from indigenous peoples. I have not had enough experience to comprehend ancient practices fully. No nonnative person I know has. Indeed, many Eskimos and Indians are no more apt to know their traditions well than an average American citizen is to know American cultural history well. During the years I lived in Alaska, there was little indigenous literature available in English or in the native languages. There were great storytellers in the twenty languages still in use among traditional people, but the literature of village people was just beginning to blossom; its beauty and mystery and truth were already apparent though not yet abundant. What I am drawing on here, in the face of our own need to create a sustainable culture, is what I heard and saw while traveling as director of bilingual education for the state and as a wandering director of the Alaska Humanities Forum, our state-based humanities program. In both cases I was trying—am still trying—to understand the differences between indigenous cultures and our own and to discover what we might learn from our indigenous relatives.

A good anthropologist might once have described an indigenous people as a group defined by its members' common culture and common territory. In recent years the definition has expanded to include knowledge systems. My own use of "indigenous" is looser. All I mean by the term is that such folks seem at ease in their place and their tradition, and that ease seems to stem in part from their longevity in that place and that tradition and in part from their development of a language that allows them to recognize the place and subsist in it along with its other animals, plants, spirits, and geologic forms. The place, the tradition, and the language have generated a kind of wisdom, which is greater than knowledge. Thus not only is a contemporary Yup'ik elder indigenous, but so are Confucius and Heraclitus, Lao Tzu and Empedocles. They are all our elders, regardless of the traditions they and we represent. I know some contemporary farmers whose wisdom makes them my elders. And I have been fortunate to know Austin Hammond, Martha Demientieff, Nora Dauenhauer, Elsie Mather, Oscar Kawagli, Eliza Jones, Rachel Craig, and other Alaskan native people, including former students like Archie and Pooch and Andy who, though much younger, were wise with the wisdom of the land and the tradition they were part of. Throughout this text I am picking the brains, and I hope the spirits, of my elders.

My personal experiences have allowed me to glimpse some tantalizing elements of indigenous life. I lived for more than a quarter century in Alaska and was fortunate during that time to visit often in Indian, Eskimo, and Aleut villages. I was there in times of grief and celebration, in blizzards and summer rains and long days of bright sun, in times of harvest and of holing up. The people of those traditions live successfully in climates and conditions most others would see as harsh if not hostile. Over millennia they developed sophisticated, complex, integrated, and meaningful cultures, yet in recent years their cultural integrity has suffered enormous onslaughts. The symptoms of their shock and pain are well known and easy to see: suicide, alcoholism, family violence, all on a scale unknown in years past. We have to acknowledge that our American educational systems, churches, military, industries, and commerce have all been part of what Ivan Illich calls "a five-hundred year war on subsistence."[5] There is no justifiable defense for the cultural devastation, ethnocide, and even efforts at genocide that Western institutions have wrought since Europeans' arrival on these shores.

Genocide? If that seems like a bleeding-heart exaggeration, consider General William T. Sherman's telegram to General Ulysses S. Grant after the Fetterman debacle at Fort Phil Kearny, near Story, Wyoming, in 1866. The fort's mission was to stop traffic headed for Montana on the Bozeman Trail, a trail that had been forbidden by treaty with the Sioux. Instead it let them pass, protecting settlers headed north, who plundered the resources of the country as they went. Inexperienced and contemptuous of "primitive" Indians, Captain Fetterman, a West Pointer stationed at Fort Phil Kearny, had once bragged that with eighty men he could ride through the whole Sioux nation. On December 21, he and eighty-one other soldiers were pursuing Sioux warriors with the intent to kill as many as possible. Fetterman and his troops, riding against orders, chased a handful of warriors over a ridge they had been forbidden to cross. Out of sight of the fort, they confronted Red Cloud and overwhelming numbers of Indians, who killed them all. They had been decoyed into a neat trap set by an unschooled warrior wiser and more knowledgeable than they about the country and the tactics it offered.[6]

On December 28, 1866, Sherman wired Grant, "I do not understand how the massacre of Colonel Fetterman's party could have been so complete. We must act with vindictive earnestness against the Sioux, *even to their extermination, men, women, and children.* Nothing less will reach the root of this case."[7] The italics, of course, are mine. Sherman didn't think there was anything especially unusual or noteworthy in the language, and he couldn't have transmitted that in Morse code if he had. Reservations and boarding schools would seem to have taken a more benign course than Sherman's troops, working for assimilation of Indian children into mainstream American culture, but the purposeful result was the death of the Indians' tribal culture, the end of Indians— ethnocide, some would say, but not genocide.

And yet, despite these peoples' losses, their indigenous view of Nature, the complexity of their relationships, and the courtesy systems they have created for dealing with one another and with all other creatures provide a complicated, elaborate, but useful model worth exploring. Even in the most acculturated remnants of many tribes, there is a core of integrity. Sometimes it seems only a vestigial remnant of internal coherence—a unity of worldview or spirit that the fragmented American culture has not had since leadership was vested in a landed gentry along the East Coast, perhaps not seen in the West since Julius and Augustus imposed it briefly two thousand years ago.

That integrity is reflected in the unity of the humanities, sciences, mores, and daily life that one can yet find in small Alaskan villages. In many villages still, there are no dancers, as we recognize dancers, for everyone dances. There are no artisans, as we recognize artisans, for everyone is an artisan. All the women in Tununuk make baskets. The baskets are neither art nor artifacts but utensils. Though one or two women are known to make exceptionally fine baskets and some make extras for sale in museums and gift shops around the state, there is little distinction between this periodic activity and other periodic activities such as berry picking and cooking. There are also storytellers, who incorporate everything the people need to know in their stories. They are not stories in the sense of Western literature but tales that include what all people might hope for in their own literature: survival skills, history, biology, geography, animal wisdom and psychology, entertainment, theology, ethics, and art—all rolled into a story told to the whole village, accomplishing an equal and universal education that mainstream American culture has yet to achieve. What distinguishes those stories from what is conventionally called literature is not that they are less literary or that they are oral but that they are seen as essential to the life of the people, while in nonnative cultures many think that literature can be set aside or ignored. Stories are seen as irrelevant to commerce, politics, industry, or science unless they can make money. Best sellers that can be transformed into movies are the exception that proves the rule. But I cannot imagine a school curriculum in traditional Eskimo culture that would segregate history, literature, art, religion, science. They are so closely related one might hear a murmur: such knowing is all one. When we separate these things, we do not reflect any greater critical or analytic skill but merely our fragmentation and disintegration, our inability to do for ourselves and to be whole persons.

Cultures have persisted for thousands of years without agriculture, without industrializing, without banking, and without literacy, but there has never been a culture, so far as we know, without music, stories, and poems. Nor has there ever been a long-term culture without an educational enterprise, and that education has always moved in two directions: back, to learn the stories of the past and get them right, and forward, to transcend the past and create a more viable future. And always, from the first campfire conversation till just recently, that education has covered the range of what we now call the humanities, the sciences, and the arts. The goal of that education—until the last few decades, when the goal has shifted to job skills—has always been wisdom: intelligence, intuition, information, knowledge, and keen insight put in service to the community to enable "the people" to survive. (And, in indigenous cultures, "the people" includes four-leggeds and bird people as well as two-leggeds.)

From earliest days, one consuming human desire has been to know why things are the way they are, and our greatest stories have tried to account for the primary "why" and "what" questions: Why is there evil? Why do the innocent suffer? Why are we alive? What is our purpose here? What is the meaning of our lives? No matter our culture, our knowledge, or our educational system, we have rarely been willing to settle for knowing simply how things work. At least till now. There now is an inclination among academics to believe that the humanities are about books, that they are dependent on the written record and therefore could not have existed before literacy. Our own National Endowment for the Humanities has been known to promulgate that very notion. The simplest insights of anthropology and even casual conversations with folks who do not have advanced degrees indicate that nothing could be further from the truth or more arrogant. Contrary to the claims of academe and the NEH, important as books have become to the humanities, the humanities did not begin with books, and they will not end when the last book has been tossed aside.

Just as there has never been a lasting culture without the humanities, there has never been a culture without a skilled application of the scientific method. The abilities to observe accurately, to create hypotheses, to test and retest those hypotheses over time, and to create meaning around our observations, tests, and the facts that grow from them have been in place since the first hunter in the world took a shortcut to attempt to intercept game. He soon knew whether his hypothesis was right or wrong, and next time he could reframe his mental arguments about where his quarry might go. Indigenous agricultural experiments with plant and animal genetics, as long as nine or ten thousand years ago, and their ancient recognition that biodiversity was critical to sustainability, support the same notion. To believe that the scientific method was invented or systematized by Newton and Bacon is to misread history by about two and a half million years.

The arts, despite their role as a frill in our contemporary culture's school programs, are always just as primary as the sciences and the humanities in the longest-lasting cultures. I have seen exquisitely carved and engraved ivory oarlocks from an Eskimo skin boat. Neither the carving nor the engraving was essential to the function of the oarlock, which would have worked just as well if plainly or even crudely carved. But they were essential to a hunter's profound aesthetic sense, integral to his understanding of himself and his most meaningful work, a means to honor the animals he hunted and to become the kind of hunter he wanted to be. The world's most ancient pottery and basketry attest to the same impulse to understand and create beauty, and at levels one hundred thousand years back, we find small, perfectly executed carvings, some naturalistic, some stylized, almost abstract. Some of those carvings are made of bone, which is hard enough, and some of stone, which may be harder still. They speak of highly developed aesthetic senses, patient sciences, and technologies created with time, intelligence, and muscle, all put in service of art and subsistence. Why? Because we cannot live long without beauty.

Indeed, beauty is essential to sustainability, according to agrarian and plant breeder David Podoll. I asked him what his criteria were for selecting seeds. "If you walked farmers out into a field of trial plots of different strains of wheat or oats, emmer or millet, and asked them to pick one," David told me, "and then asked why they picked it, the most common first response would be to consider, shrug, and say, 'Oh, it's just beautiful.'" David said it took him a while to learn about the importance of beauty. Now he believes that "all the most durable qualities of a plant, or of a sustainable food system, follow on from beauty. The criteria for sustainability [have] to include beauty." His comments give us some new ideas about the nature of beauty and what might constitute it. And about the criteria for sustainability.

Once, while working with his brother Daniel on saving seeds from a particular kind of squash, David also learned that plants have "spirit." The color of the seeds was washed out, the skin didn't seem right, they were tough, and the seeds were hard to get at. The barn the brothers work in got pretty quiet. "Those squash just had a sour spirit," David said. But then they moved on to another strain, and the weight was so hefty, the color so bright, the fruit so firm, the seeds so white and so easily freed that the whole barn brightened up. He and Daniel were talking and laughing, and the seed saving went along briskly and pleasantly. "Those plants just had a better spirit," David told me with a grin, "and it affected the whole place." In a conversation in David's living room, Nebraska farmer and organic certifier Tom Tomas unwittingly echoed David's view. "The measure of healthy soil, of good health, is beauty," he said, and then added, "It means that you have found the balance between raising food and doing no harm to the natural environment." David's and Tom's views are neither science nor religion. They are more important: they are indigenous.

Another characteristic of all those means of learning for long-lived cultures is that they are not as fragmented and uncommunicative across disciplines as they have become in more specialized cultures. Indeed, often the same person is the basket maker, potter, quill worker, food and medicine gatherer, and dancer; or the carver, storyteller, tool and equipment maker, hunter, and dancer. Each one understands the science, technology, aesthetics, and lore that allows proficiency in every arena. Such learning is not less complex; it is as difficult as any other, and traditionally it was undertaken without tools and equipment for study. It includes metaphysics along with other learning. Marie Meade, a Yup'ik Eskimo from the Kuskokwim villages, reveals through the words of the people that making masks, presenting them in dances, drumming, and singing are all "our way of making prayer."[8]

For the longest-surviving cultures, the sciences, the humanities, and the arts were shot through with the sacred. Nature and the sacred, wisdom and the sacred, were inseparably linked. Rituals for establishing those ties were part of every child's education and every adult's daily practice. Indeed, it was intense, personal awareness of the sacred that led to the simultaneous, and equally important, creation of the arts, the humanities, and the sciences. In our time the sacred has come uncoupled from wisdom; wisdom uncoupled from knowledge; knowledge unhooked from information; information unhooked from facts; facts disconnected from data; data disassociated from firsthand observation and experience. Our culture seems to have forsaken wisdom in favor of all the latter—at a time when wisdom is our greatest need and would be its greatest asset. Personal observation and experience are now relegated to "anecdotal information," a very diminished status compared to that of data. That is an odd stance, since the best science is still rooted in observation and experience, either in the lab or in the field. "Where is the wisdom we have lost in knowledge?" T. S. Eliot asks. "Where is the knowledge we have lost in information?"[9] It may be that in the long run (where intelligence and wisdom always lie), we will discover that the great contemporary tragedy is the exaltation of information. One can easily discern the distinctions Eliot makes: Information leads to a quick killing on Wall Street; knowledge leads to a sound economy of the kind professional economists define for us, in which the economy can be deemed "booming" though millions starve and more millions lack food security. Wisdom leads us to a healthy culture that provides subsistence sufficient for the whole community without crippling the next generation's chance to subsist as well.

There has never been a culture that lasted for long while ignoring its land base, extracting so much from the land's capacity to nourish that it could not regain its own composure and regenerate itself. It is a truism that everything comes from the land and goes back to it, and the land comes first. Sustainable traditional cultures were too small and too mobile to utterly consume all their resources. Nevertheless, they paid closer attention to their resources than we have learned to do so far. In our time it is population growth and resource exploitation—not for the survival of the people, but for increased wealth among the already wealthy—that have brought us to outstrip our resources.

Long-term cultures are profoundly, rather than superficially, democratic. They have courts, but they do not have representative governments and they do not have to balance legislative and executive branches. Their kind of participation avoids a 51-49 vote in which the winner takes all and leaves the community torn, half its population aggrieved. Former chief justice Arthur Goldberg told me in 1971 that "democracy and the rule of law mean that you consent to lose," but there are indigenous cultures far longer lived than ours in which people talk things through with one another long enough that no one loses. The notion that modern democracy began during the reign of John, king of England, or with John Locke, or during the American Revolution is to misread our political history as greatly as we do the history of science.

There has never been a long-term, sustainable culture that did not keep an eye on reciprocity, balance, and harmony as matters of official public policy, incorporating constraints on the behavior of its people toward animals and plants and the world generally, as well as toward one another, so that the human life of thought and activity and the larger ecological life could all participate with least harm to the other, a kind of democracy of the biota. Those old cultures had an honest, comprehensive accounting system that did not allow for "external" costs. Nothing was considered external.

Long-term cultures developed rituals of restoration, ways to bring the exile home and to restore balance and harmony between individuals and the culture, and between the culture and the ecosystem. These, too, were matters of official policy. Where are the concerns for restoration, for bringing the exiles home, and the rituals for reestablishing balance and harmony in our public hearings about land use policies, the spread of a global economy, mergers of megacorporations, economic development, or metro sprawl? Where are these concerns in the policies of our major environmental organizations that often seem focused on driving perpetrators at least into court and further into exile? Where are these concerns in the policies of our corporations that often paint environmentalists as destructive fools, tree huggers out to impoverish neighbors and businesspeople and stop economic growth? Indigenous policies of balance and harmony do not limit fairness to the humans involved but include the entire ecosystem in which humans are also a part. One has to take the neglect of these things in official policy to be another indicator that we do not yet have a sustainable culture, do not have much understanding of Nature in spite of our science, and have lost our way in the world.

Long-lived cultures tend to be "civil" cultures, with elaborate courtesy systems in place to avoid confrontation and alienation. Among some Athapaskan peoples on the Yukon, for example, it is considered impolite to ask a direct question. While preparing for an elders' conference, Holy Cross elder Martha Demientieff told me that someone who wanted to know why the elders had done something in the past might muse, in their presence, "I wonder why they did that." An elder would be free to respond to the musing yet not compelled to answer. In contrast, the direct question "Why did you do that?" might feel like a trap or even an indictment, for the interrogative demands an answer. Eskimo linguist and writer Elsie Mather once told me that among Yup'ik people along the Kuskokwim River, disputants sometimes seek out a third party to act as go-between, so the two parties do not have to face each other in a showdown that might permanently rupture the social fabric. In Tlingit cultures, it is the uncle, not the father, who trains the young boy in hunting and fishing and the skills necessary to maintain a respectful life in both the human and wilderness societies. That way, youthful chafing or outright rebellion against discipline is not aimed toward the father, and the nuclear family can maintain decorum, explained Ellen Hays, a Tlingit from Sitka, Alaska. Similar courtesy systems extend to game, fish, and the rest of the environment, so that what we call "civility" ramifies throughout the culture and includes all of the natural world. I do not want to romanticize here by making villages seem conflict free; villages are often torn by difficulties, for example, between two large family groups, and individuals have their differences. But each culture tries to anticipate and head off such events and offers civilized systems to mend broken relationships. Modern American culture now seems to tend toward revenge, or to prefer separation and punishment—jail—for those who disrupt it or fail to follow its taboos.

Contrary to popular perceptions of "primitive savagery" and tribal warfare and violence, there has never been a long-lasting culture based on war, violence, repression, or slavery for the majority. There haven't been many cultures, long or short, without warfare, but in the longest-surviving cultures it was sporadic, usually seasonal, often more "aggressive ritual" than murder. The long-term siege was not seen as a necessary strategy. Troy is a recent event. Philip of Macedon, Alexander, Genghis Khan—all were upstarts on the land who diverged from the most ancient cultural growth. I am talking about cultures that have been around since long before the Greek polis, as well as before the Confucian flourishing of China, another relatively recent occurrence. The purpose of war among indigenous peoples was most often theft of women and trade goods and rarely—very rarely—territory. The cultures that have followed a violent or oppressive course have most often been broken apart by internal conflict or by weakness induced by too great attention to internal controls; so much of the available resources go to control that the capacity to defend against invaders is diminished. The old rule of the western movies is ironclad: there is always a faster gun. Sooner rather than later (in the definition of longevity we are using), the alpha male goes down and the oppressor becomes the oppressed, for the violent cultural center cannot hold. Two essential characteristics of a sustainable culture appear to be what we now call social justice and peace. Resorting to violence is always a clear sign of a culture at risk rather than a culture of strength.

And again, because this is perhaps the most important observation, there have been cultures that persisted for thousands of years without literacy, but there has never been a sustainable culture without healthy stories. Our real power in America, unrecognized, lies neither in our military nor in our economy but in our capacity, limited though it often feels, to tell ourselves healthy, rather than crippling, stories. What follows may help clarify the distinction between healthy and crippling stories.

Many sustainable farmers believe that sustainability is about experimenting with new agricultural methods, or old ones, and fighting the politics and power of monoculture agribusiness. I think they're writing a new story about how to farm in this world—a healthy story that will transcend and replace the story that has been drilled into our farmers' heads for fifty years by agricultural industries and land grant colleges. Their story has been that chemicals will work miracles, that bigger is better, that good farming is a business to get rich in rather than a way of living on the land—a crippling story that helped create a crippled and now crippling agri-without-culture. Living on the land has to include sound, sustainable business practices as well as intimate, personal knowledge of the land, crops, water, and livestock. Agriculture under the leadership of transnational agricultural industries is not improving. After all, how sustainable is a farming system in which sons or daughters cannot, or will not, farm? How sustainable is an agricultural system whose toxic runoff pollutes the wells and city systems of its urban neighbors who are the market for its produce? How sustainable is an agriculture whose pesticides and herbicides kill its workers?

Our culture has insisted for a couple of hundred years now that the primary unit of value is the individual—an individual's ability to realize himself (herself, alas, didn't figure into this until recently) regardless of consequences to others, to the larger culture, or to the land. That is nonsustainable nonsense. There has never been a sustained culture in which the individual exceeded the community in value. So now we need to create a story that puts individual freedom back into responsibility to the community, a community embedded in, surrounded and sustained by, larger social and ecological communities. Indeed, in the oldest cultures, recognition of the unity of the human with Nature was so clear that both were part of the same social and even linguistic order. If we can reintegrate individual freedom into the community, individuals will rediscover the freedom that can be found only in community, and we will have a healthier story to tell ourselves and our children.

In our homes, our schools, and often our churches, children have been told for years that Western culture is the best in the world, that God has blessed us above all others and loves us most, and that Anglo males are the strongest, bravest, best, and smartest humans in the world, real can-doers who can solve any problem regardless of scale. That's nonsense, of course, another crippling story that has crippled or destroyed the lives of millions of non–Anglo American citizens and many Anglo women as well. It's a story that also diminishes, and ultimately demeans and degrades, Anglo males. Most of us are pretty much like everybody else of whatever color or country, struggling to make our way in an unsustainable system. The new, healthy story of how to live together will recognize ethnic, linguistic, and gender diversity as differing forms of biodiversity and as values equal to biodiversity. That healthy story will also acknowledge that, whether rich or poor, if we are trapped in an unsustainable system, we will all go down. There is no way that wealth will buy us either safety or a way out.

Science cannot save us for three reasons. First, science is expensive, and now it works for the highest bidder. The highest bidders, even in universities, are often transnational corporations whose object is not sustainability but short-term shareholder profit. Second, though science does ask important questions, it does not even want to ask some of the questions critical to a sustainable culture. Those questions are basic ethical, moral questions of our relationships, questions that science eschews as too subjective to be amenable to hard scientific inquiry. The healthy questions that lead to healthy stories from our own history have to do with democracy, equality, opportunity, compassion, and freedom. Information, sometimes critical information, comes from science; wisdom for our living in the world comes from other stories. Third, even science can be found napping, lacking the foresight and policy influence to command sufficient attention to direct research to appropriate questions. We may well pass a point of no return before we can recognize it and thus fail to survive the impact of climate change, for example, or the chemical pollution of our environment. Scientists, alas, are no more inclined than the rest of us (and perhaps less inclined) to see the big picture or take the long view, to examine the thousand-year implications of their activities.

The new global economy is a white whale called progress; commerce and industry are our Ahab in mad pursuit; and we are all the *Pequod*'s crew. Who will be our Ishmael, left to tell the tale? My bet is that it will be someone Yup'ik. I remember Robert Clark, a Yup'ik Eskimo from Bristol Bay, saying that although he had a powerboat, and used it, he continued to practice using his kayak because "there will come a time when you and all your machines will be gone again. We will be back to old ways. Then I will be ready." Clearly, Robert did not believe that we ignorant white savages would be around for the long haul.

Our economics, social life, politics, and schools have also insisted that having more toys is better than having fewer toys; that buying stuff is good for us; that we have to keep up with (or, rather, exceed) others in our consumption; that a high-paying job can take the place of meaningful work; that low-paying, meaningless jobs that demean our humanity are better than none, and we should be grateful for them because they will turn us into decent citizens; and that a free market has the same beneficent powers as a just god. But capitalism rests ultimately not on innovation or entrepreneurship or brains or even a free market—those are just stories we Americans like to tell ourselves because they make those who are successful look good. At its base, industrial capitalism's success rests on exploitation of resources (consider the state of the environment, and the ways industrial polluters have found to externalize costs), racism (consider agriculture's exploitation of members of ethnic minorities for picking and industries' use of them in food processing plants), child abuse (remember those children of the Industrial Revolution left to sleep under the bridges after years of standing at looms pumping the treadles until they grew into their misshapen and deformed teenage bodies, at which time they were fired—and consider that America had laws to protect dogs before it had laws to protect children), sexism (consider the use of women in maquiladoras and "free trade zones" today), and war (consider the lingering impact of the Great Depression until World War II and the subsequent persistent development of war industries that helped generate a robust economy, an economy in which military production and weapons are still so important—and growing rapidly again—that dismantling it would trigger massive unemployment and economic collapse). Weapons are the United States' biggest commercial export. The Enlightenment has some very dark corners.

But as much as, or perhaps even more than, all these, contemporary capitalism rests on consumption: government and corporate consumption of resources, technology, and scientific research, and citizen consumption of market goods. As big a business as any in our culture is an unexamined and unchecked advertising industry designed to get us to consume, another crippling story rather than a healing one. We are asked to consume not only material goods but ideas, policies, whole worldviews that are presented with all the persuasive skills and battering psychological hype that can be bought. We are under assault, being laid siege by hype: corporate hype, political hype, military hype, educational hype, commercial hype. The U.S. government would have us believe that we can fight terror by getting back to our shopping as soon as possible, even if it's limited to duct tape. As our civil rights have declined in recent years, freedom has come to mean the freedom to choose among sixteen brand names of one product. When the market was local and what was made (that is, manufactured, made by hand) was essential goods, consumption and production were closer, and the goods people bought, or bartered for, served their work and their lives and often fed their aesthetic senses as well. Now the majority of what the culture produces is nonessential, aesthetically unsatisfying frippery and frills. One result of this is that American consumers frequently spend more money on the package than on the product. No wonder we have become participants in an unsustainable culture that is eating itself, and us, out of house and home.

Our typically American romance with growth has resulted in a mandate for agriculture that has destroyed our small farms and small towns and raised havoc with our topsoil. For fifty years now, that cultural mandate has been get big or get out. Those who farm have heard it as a continuous din from corporations, land grant universities, and the government, all institutions we once believed in and relied upon. They have been telling us a story that cannot sustain us for the long term and has not served us well to date. It has resulted in growths we do not want—in our wives', mothers', sisters', and daughters' breasts and our husbands', fathers', brothers', and sons' colons and prostates. This is the harvest of a culture so bent on growth with all possible speed that it will pour a hundred thousand chemicals into the earth and atmosphere, into our lakes, groundwater, and oceans, before it has a clue about the long-term effects of a single one of them.[10]

These are just some of the unhealthy stories we have been telling ourselves. They have been killing us, literally, but they are so powerful and we so want to believe them that we have ignored what they have done to foreshorten our future.

One thing I learned from talking with farmers and touring farms the last few years is that agriculture is one of the sites where the environment and social justice intersect. So, too, science and the humanities, information and the wisdom inherent in stories meet—or fail to meet—in profound ways on every farm. If the first rule of sustainability is that all things are related, then all these elements are essential to a sustainable culture, critical not only to the larger urban culture but to agriculture.

So we've got to find a new story to tell ourselves about sustainability and the land, and we've got to find it fast. Farmers are falling like flies, too many by their own hand. Too many bought the myths of chemical miracles and "big iron" machinery that have imprisoned them in an unsustainable life of debt and failure. They too often lose the race against debt, volatile prices and markets, crop disease, and the many insect species that respond more rapidly to changing circumstances than humans do and outstrip our own inventive capacities. Our chemical pesticides, herbicides, and antibiotics cannot keep pace. That is evidence, one has to assume, that soybean aphids are simply more intelligent than the soybeans they crave, and that both the aphids and the soybeans are inherently more intelligent species than the humans who are bent on destroying them. Genetic modifications and chemicals are only patches, short-term fixes that so far have proven more destructive than constructive, although there has not, as yet, been either interest or time enough to do the studies of their long-term effects.

We have known of the ties between landscape and human and animal health for many years. The latter two depend on the first. Our rural social scientists have known for at least half a century that when land health declines, our economies decline, rural towns and their populations and markets shrink, and churches, schools, colleges, banks, and businesses cut back and close. Psychologists and social workers know that when the land's health declines, stress increases; depression, alcoholism, child abuse, and spouse abuse also increase; whole populations are uprooted; and intellectual and spiritual resources are depleted. Botanists, agricultural scientists, and chemists have known about the links between the health of the soil and the health of everything else for even longer. In the Western tradition, Lucretius, Hesiod, Cato the Elder, Caesar, Cicero, and Seneca knew it a couple thousand years ago. On the Eastern side, Confucius, Lao Tzu, and the Buddha knew it too. And indigenous cultures knew it millennia before all of them. Why should it be so hard for us to understand today? Everything that we know and cherish depends on the health of the land, and the physical basis of that is now, again, clear and absolute. When agricultural systems are damaged by biological simplification, environmental costs escalate and the human economic, personal, and social costs skyrocket. Confident in our own superior knowledge, we have forgotten—or ignored—the insight that everything we do affects everything else. Now our land, our communities, and our human spirits are all in need of healing.

In 2007 Grand Forks, North Dakota, celebrated the tenth anniversary of the great flood that brought the town to the front pages of newspapers across the nation, looking like World War II photos of Stuttgart. The Red River valley of western Minnesota and eastern North Dakota is mostly flat, an ideal landscape for row crops, so corn, soybeans, and sugar beets have predominated for years with their concomitant chemical inputs, lack of ground cover for much of the season, erosion, and flooding. It was this row crop agriculture on vulnerable lands that caused Grand Forks to sink beneath the river. Our land grant colleges and universities are implicated in this; they have been telling and retelling that story of bigger is better and quick chemical fixes and high interest loans from the hucksters, and we need to help them find a better story to tell.

I'm not a practicing farmer and have no desire to be one, but I've had enormous respect for farmers since I came to know my German immigrant grandfather when I was a little boy. He was a farmer in Iowa, a dairyman who delivered milk in town and raised sheep, pigs, horses, chickens, and a large garden—a significant portion of which went to gladiolas. One of the images that lingers from my earliest childhood is of Grandpa walking out in the dusk after supper, crossing the lane that led from the county road into the farm, ducking under a row of pines to stand in a small plowed field of rich, black Iowa loam lying deep on the field, reaching down to pick up the microorganism-saturated dirt that crumbled so easily in his hand, and spilling it into the evening breeze. He was not gloating over his ownership or rejoicing in his ability to dominate the land—a notion anyone who has ever farmed or ranched finds incomprehensibly idiotic—but was simply feeling self-possessed, at peaceful ease after a day of hard work. Like any craftsman savoring the material he works with and loves, he was a good man who simply loved good soil.

When I was nine I worked all summer for my uncle Bernard Eden in Calamus, Iowa, cultivating corn and beans. I began on a little gray Ford-Ferguson tractor, but during the summer, as the corn and beans got taller and my driving skills improved, I graduated to a much larger—or so it seemed to me; it was the one Uncle Bernard drove—Allis-Chalmers. I learned something about work then, about long hours, about the way the land demanded attention and care if it was to yield its bounty, and, I must admit, about the pure privilege of consumption on a Saturday night in town with four well-earned dollars in my pocket—a whole week's wages after board and room—exactly enough to buy a pocketknife I coveted.

I admired both Grandpa and Uncle Bernard, and I can support sustainable farming efforts in part because I appreciated their good efforts, admired both their knowledge of and love for the land and its critters, their capacity for hard work, their generosity, and their rock hard integrity—characteristics I often recognize in those ranchers and farmers I've come to know in the West and the Midwest, and in other folks I know like them.

I have also supported other elements necessary to a sustainable culture, serving for years on the board of the Center for Children and Families in Anchorage, trying with colleagues and friends to get a grip on child abuse. I have spent time developing public programs in the humanities in several states on the assumption that the humanities are essential to indigenous cultures that have sustained themselves far longer than Western civilization has to date, hoping that bringing the humanities to bear on issues of serious public concern will help to sustain my culture. I have served on arts councils for the same reasons, on commissions looking into public health issues, on committees trying to find ways to reduce racism, in environmental organizations; I have consulted on poverty, and tried to write my own poems and essays. All of these I see as fitting parts of a larger whole, the effort to develop a sustainable culture built on a healthy spirituality and an awareness of our subsistence base. I believe that the soil of that spirituality, the indispensable element that nourishes a healthy spirituality, lies less in religions than in language—in a culture's best stories, songs, and poems—some of which do, indeed, come from its religions.

Perhaps I seem a hopeless romantic, and naive to boot. But the alternative these days seems to be cynicism. I will happily trade cynicism in for a clear view of how the world actually works and will cling to hope, wherever I find it, that a vision of a sustainable, compassionate, respectful culture can yet be realized. We can find our way toward that vision from the models of indigenous peoples that are still at hand, indeed, in all of our own inner spiritual and intellectual resources.

I've been committing myself to one quixotic cause or another all my life, but now I see sustainable agriculture as part of a larger cause that, if it turns out quixotic, will spell catastrophe. So I'm willing to work on sustainable farming, and these other fronts as well, in the hope that they all move toward creating a sustainable culture. What we learn from truly long-lived cultures is that if we desire sustainability, we will focus more on language, music, and stories than on technology and economic development. One virtue of such a view is that science, too, can tell us useful stories that provide us information and knowledge needed to create health—but those stories must take their place alongside the stories that come to us from the humanities and the arts that provide us with the wisdom necessary to create a culture that can sustain us all.

Evidence is mounting that indigenous knowledge may be not a romantic but a realistic route to the future, even for agriculture. It is revealed in the wisdom of the ancient practices of peasant farmers in Latin America. Nearly three decades ago, archaeologists uncovered more than 170,000 hectares of Indian farms in use three thousand years ago. Those farms consisted of "raised fields of seasonally-flooded lands in savannas and in highland basins," writes Miguel A. Altieri of the University of California, Berkeley. "These platforms of soil, surrounded by ditches filled with water, were able to produce bumper crops despite floods, droughts, and the killing frost common at altitudes of nearly 4,000 meters." In 1984 Peru's state agencies and nongovernmental organizations began to assist farmers in recreating those ancient beds. They found that the raised beds and their water-filled ditches helped to moderate temperatures and extended the growing season significantly. This method led to higher productivity than did the use of chemical fertilizers. In the Puno district, conventional methods average one to four tons of potatoes per hectare per year. In the Huatta district, contemporary farmers who implement raised beds have produced a sustained yield of eight to fourteen tons of potatoes per hectare per year.[11]

Altieri reports on another ancient system of terraces that was recreated in Peru, "Crop yields have improved significantly. For example, potato yields went from 5t/ha [tons per hectare] to 8t/ha and oca yields jumped from 3 to 8t/ha." All this by following a "conventional" practice in use a millennium before the ancient Greeks. At the time of Altieri's report, there were "1,124 hectares of terraces and 173 hectares of drainage and infiltration canals." The results: "Enhanced crop production, fattening of cattle and raising of alpaca for wool have increased the income of families from an average of $108 per year in 1983 to more than $500 today [1994]." Altieri documents similar successes ranging from organic farming in the Andes and other places to agroecological approaches similar to those used in Brazil.[12]

All those developments, and our best hope for achieving a sustainable culture, stem directly from getting back to basics: rediscovering indigenous intelligence, eschewing mainstream American culture's bent toward arrogance, and combining that new-old wisdom with our own knowledge.

Habitat for a Sustainable Culture

PERHAPS SUSTAINABILITY IS always more complicated than it seems. We devote our attention to recycling and discover that our public health is becoming public obesity and public diabetes. So we work to restore public health, only to discover that another element of sustainability has now slipped through our fingers, slick as fresh liver that we have to scramble across the kitchen to pick up. Indeed, I have spent time working on health issues in our town and, from there, gone to a meeting of about thirty folks in town forthrightly called the "racism committee." For several years they got together once a month to talk about what they could do to reduce the racism in a town steeped in a century and a half of it. Further, the difference between the wealthy and the poor in our community is grimly visible; our women's shelter is fully occupied with battered and frightened spouses; and gay men and lesbians move about freely only because of their courage. At some level, good health or not, the community is not a community but a collection of fractured enclaves—racial, ethnic, economic, gendered, religious—and the fractures are deep enough that we may never be able to knit them. If we can't, our town is not sustainable but will come apart at the seams.

The Three-Legged Stool

There is a familiar metaphor one hears in workshops on sustainability these days. It is the image of sustainability as a three-legged stool. The three legs are the environment, our social life together, and the economy. This is meant to represent a habitat for a sustainable culture. The image is popular—and poisonous to the cause of sustainability.

When we think about sustainability, we commonly think first about the natural environment. The initial interest of wilderness advocates was in protecting habitat for the sake of humans, at first for our aesthetic pleasure and our psychological or spiritual health. A bit later, as our understanding increased and the extent of destruction grew clearer, we became concerned for our physical health. Now we are gradually moving beyond our anthropocentric concerns and including the health and well-being of other creatures, recognizing that we are all related and that the health of all, both psychological and physical, is interconnected: the death of the snail darter or the spotted owl is but a symptom of our own failing human health, the dying cell that presages a dying organ that will destroy the larger body.

The source of our environmental concerns is easy to see. Driving over Snoqualmie Pass in the early evening, we can see the clear-cut hills, and the runoff from the rain covers them like a sugar glaze, glistening in the dusk as the water cascades over them to flood cities in the lower reaches. The logging roads, zigzagging across the slash, are even uglier—scars that will not heal in my lifetime nor my children's, nor their children's. In the Midwest we cross old bridges, look at what appears to be beautiful habitat for fish, and remember the warnings that we should not eat them because toxic materials accrue inside their bodies and will also accumulate in our own. Even on these gentle prairie slopes, the signs of erosion are too often clear. No matter where we live or travel, we can all point to environments that have been devastated, and we ache to stop the damage and make them new again and whole. That is one aspect of the habitat for sustainability requiring our immediate care.

More recently, the desire to create sustainable communities has led us to consider social health as a second habitat that requires our attention, the second leg of the stool. One of the fundamental assumptions of our efforts to achieve sustainability is that we are all in this together, uncomfortable as that may sometimes make us. "For nature demands that all things should be right and harmonious and consistent with it-self and therefore with each other," said Cicero, the old Roman senator, more than two thousand years ago. In business transactions we "are obliged," he says, to tell a customer everything that would be useful for him to know, remembering "that nature has joined mankind together in one community. . . . So everyone ought to have the same purpose: to identify the interest of each with the interest of all."[1] How much we have forgotten since Cicero's day. We don't get far without the involvement and aid of others. This, too, has parallels in our agriculture.

Geneticists' selection of plant or animal characteristics, described by Wes Jackson in a talk on genomes at Wartburg College in Iowa—se-lecting for a single plant virtue while ignoring others—has its equiva-lent in our social environment. It is called ethnic cleansing. The former represents a dead end for biology; the latter, for society.

Language diversity is as critical to a sustainable culture as biologi-cal diversity is to the rest of creation. Worldview, I think, is less depen-dent upon sight than on language, and when a language is lost, the way that language can see the world is lost too. It lies beyond revelation by other words. Consider the sense of loss when you know your children and grandchildren will never see the world the way you do, and you know that is not only because the world has changed so much but also because they do not have the vocabulary to see it with. You know the perceptual capacity you have is being extinguished in your children, dimming their vision. No wonder American Indian people often feel disoriented, alienated. They are aliens in the land of their birth, blinded by the loss of words. If justice does not require that the dominant cul-ture maintain languages different from its own, then compassion re-quires it, and ultimately wisdom, the mother of compassion, requires it, for each language is a window on the world. No one culture or language can provide us with a comprehensive, all-seeing worldview. We need all the worldviews that many languages make available to us, all the pos-sible ways of seeing the world that many languages allow. The world now holds about 6,700 languages. Linguists estimate that by the end of this century one-half of them will be gone. We need the diversity of languages for the possibility of their healing insight and their expansive connections to other worldviews, just as we need rainforests for their undiscovered healing medicines. The loss of native languages is a loss beyond words, and it imperils not only their speakers but all humans. Indigenous peoples around the world have the same right to keep and use their language as English speakers have to keep and use English. Yet the toxic pattern of "English only" that the United States, particularly our education system, has followed is as brutal as gladiatorial combat.

Recent world events have shown again that the old colonial days are dead, and the so-called new world order of the mid-1990s was just colonialism behind a new mask. In the social world, as in the natural environment, whatever creates imbalance creates an unsustainable world—whether that imbalance is racial, economic, or educational, and whether we think of it in terms of consumption, access to real information, or opportunity. So our social life—all those elements of work and love and family and ritual and celebration and grief, and all the babble of voices in their myriad tongues—is as critical to creating a sustainable, regenerative world as is our care for the natural environment. One of the virtues of the image of the three-legged stool—and the concept of sustainability—is that it reminds those of us who love wilderness that the social side is also critical to a healthy habitat for a sustainable culture.

Now consider the third leg of the stool: the economy. As it stands in the current metaphor, this third leg is separate from the environmental leg and from the leg that represents society at large. But to separate the economy from the environment and the society grants it an independence it neither has nor deserves and lends it an authority that exceeds its real power. Every economy has an impact on the environment—even the highest-technology, theoretically pollution-free economies require energy and leave toxic detritus behind. And any economic system that exacerbates the gap between the wealthy few and the impoverished many makes a world that cannot sustain itself. Separating the economy from the society and the environment is fatal to the creation of a sustainable culture. It may lead us to make allowances for commerce and industry, allowances that inevitably result in harm to our natural and social environments. We have already made excuses for them too often, to the harm of the earth and ourselves. The economy is a human creation, a social phenomenon with powerful, most often destructive impacts on social and environmental health. The economy belongs not in a separate category but as one of many social indicators. Despite the claims of economic determinists, the economy is merely a subhead we have to work on along with racism, education, the arts, and cultural opportunities.

Let me be clear here. I am not trying to diminish the importance of the economy or of jobs, for healthy societies generally have healthy economies. But a growing, even booming, economy, we have learned to our confusion and sorrow, does not necessarily lead to a healthy society. Witness America from 2000 to the present. Any well-paid economist will tell you that the economy is doing fine. Any farmer, migrant laborer, or factory worker will tell you that the economy has gone south. Indeed there is evidence that a wealthy society is less healthy for both the environment and the local culture than a poor one and that even poverty-stricken economies can sustain cultures that last for a thousand generations and create astonishing works of art, thought, and craftsmanship.

If we have a strong, flourishing economy but we abuse our children and assault our spouses, where are we? We are a far cry from a sustainable community. During my years on the Anchorage Child Abuse Board, I saw pictures of a baby covered with cigarette burn marks, all laid out across the tiny back in deliberate patterns, as if tattooed, and pictures of the pulped faces of women attacked by their husbands. I know women who have been hit; I have seen the blackened swelling. Those who care about the environment cannot dismiss these as social issues. If humans are part of Nature, as our environmental rhetoric insists, such scenes are a blight not only on our social life but on the landscape—a blight on the landscape just as surely as a clear-cut is—and are as fatal to a sustainable culture as any economic collapse. Too often those of us who love the wilderness have used rape as a metaphor for the loss of trees or whales and ignored the rape that is not a metaphor.

Such abuses are beyond the reach of the economy and have nothing to do with income or education or social class. They are immune to the amenities that wealth affords, and reducing poverty won't touch them. Lawyers do it and laborers do it; doctors do it and trash collectors do it; college professors and those who quit school in fifth grade do it. And those of us who do not do it know that somewhere down inside ourselves, hiding in the shadow inside us all, is a self in us that could do it too. If humans are a part of Nature and not something outside it, an abuse victim falls like a felled tree.

When we separate the environment, the society, and the economy into three legs, we fall into the trap that our science and philosophy have laid for us for four hundred years. We separate humans from Nature, and the economy from both as if it were something beyond human influence, equal to the natural and social environments in its independent authority. But the economy is a human cultural construct, never independent from either the environment or society, and only a part of those other aspects of the habitat for a sustainable culture—not separate, and nowhere near equal. Further, when we separate a concept like sustainability into three parts, we reduce something that is incredibly complex to something that is intolerably simple. The natural world and our lives in it are far too complex for that image. To come closer to reality, we have to reintegrate all three of those metaphorical legs into a single leg.

The One-Legged Stool

The idea of a one-legged stool may appear foolish to our contemporary minds, but it is an image that will serve us well here. It is the image of my grandfather's milking stool. It had only one leg. The balance was provided by Grandpa's sitting on it. Without that it would not stand up but would lie on its side where he tossed it in the corner with a smooth swing as he stood up from milking the last cow.

Grandpa's sense of balance leads us into another inseparable environment that I think is critical to both a healthy society and healthy ecosystems. It is the first element in a sustainable culture, the first arena that needs our attention and care as we seek to create a sustainable life. It is an element entirely left out of the image of the three-legged stool, and out of most discussions of sustainability. I don't have a satisfactory name for it, but this other environment has something to do with the idea of a spiritual habitat. I am not thinking of "spiritual" in any traditional religious sense and am not limiting it to that aspect of our lives wherein lie such characteristics as compassion or empathy. This environment is intellectual, emotional, and, well, spiritual. Grandpa's sense of balance was not entirely physical; it stemmed from a right spirit as well. At the same time we work on a regenerative life for the natural and social worlds, we must work on ourselves, on setting ourselves right. In the *Hua Hu Ching*, Lao Tzu asks,

> Would you like to save the world from the degradation
>   and destruction it seems destined for?
> Then step away from the shallow mass movements and
>   quietly go to work on your own self-awareness.
>
> If you want to awaken all of humanity, then awaken
>   all of yourself.
> If you want to eliminate the suffering in the world,
>   then eliminate all that is dark and negative in
>   yourself.
> Truly, the greatest gift you have to give is that of your
>   own self-transformation.[2]

Remember the Qur'an's warning? "Allah does not change a people's lot unless they change what is in their hearts." The key first step toward the creation of a sustainable habitat for our natural and social worlds lies in the health of this third, interior, spiritual environment. It has to be the healthy, intelligent, wise, caring, internal base from which we create health for ourselves by creating health for everything else. Such an apparently altruistic impulse comes only from the internal spiritual habitat I have in mind.

Ralph Waldo Emerson, so grandly wrongheaded in so many of his grandest ideas, was also right in some. He thought that "the problem of restoring to the world original and eternal beauty, is solved by the re-demption of the soul." He adds, "The reason why the world lacks unity, and lies broken and in heaps, is, because man is disunited with himself. He cannot be a naturalist, until he satisfies all the demands of the spir-it."[3] Emerson implies that if we tended to this personal, interior life first, the others would be taken care of automatically. In this he echoes his friend Thomas Carlyle, with whom he surely discussed such matters. Carlyle writes, "To reform a world, to reform a nation, no wise man will undertake; all but foolish men know that the only solid, though a far slower reformation, is what each begins and perfects in himself."[4] The world needs all of us involved in this process of self-transformation. None of us are exempt, none are unnecessary to the work that needs to be done.

Though we may not be able to measure our inner intellectual, emo-tional, or spiritual life, we can see measurable expressions of that life. The health, harmony, balance, beauty, and intelligence—or lack of these—in our towns and our countryside are a direct outgrowth of the health of our souls. If what comes out of our cities and flows into our rivers is garbage, that is one unhappy measure of the state of our spiri-tual life, a sign that there is still a little garbage in us.

The heart of sustainability for agriculture, according to Wes Jack-son, lies in the state of the nutrient supply in the soil. Take that last phrase metaphorically and it becomes the base for the health of the natural environment and for all the other aspects of a habitat that will support a sustainable culture as well. Soil, not landscape, is the soul of all our earthly enterprises: social, economic, environmental, spiritu-al—however you want to name or categorize them. The nutrients for the social environment are reflection, thoughtfulness, empathy, com-passion, and understanding. The special nutrients of the internal envi-ronment are restraint, reflection, and deliberateness, perhaps what Buddhists would call "mindfulness" or Confucianists would call "self-cultivation." That mindfulness or self-cultivation will take account of the genius of the place, as Wes says all good agriculture does—in this case our human, interior place—and then will act deliberately in light of what it knows and acknowledge "the reality that we are billions of times more ignorant than knowledgeable."[5] When we finally get the right balance of these nutrients mixed in the soil of our souls' interior life, we will be on the way toward creating the habitat for a sustainable life. It is helpful to note that many healing rituals of indigenous peoples are aimed precisely at restoring balance and harmony within the person who is ill. The assumption is that illness stems from being out of balance with one's environment. Our own balance on the one-legged stool, derived from accepting our place in Nature rather than fighting to expand or reshape or deny it, will keep everything upright. The primary resource for the development of this area of the habitat for a sustainable culture is language, the vehicle for stories, songs, poems, and the means for exploring philosophy, theology, and even spirituality.

Yet another reason for putting all these legs (and ourselves) together is that we no longer have the luxury of working on these aspects of a healthy habitat for a sustainable culture one at a time. They are equally in need. We cannot save the whales and neglect our central plains. We cannot protect the great wilderness landscapes and ignore our topsoil. We cannot protect old growth forest and ignore the health of our spiritual lives. Neither can we retreat into our spiritual lives and ignore the trees and whales. That won't work; it won't save us. Whatever we create out of that single-minded approach will be but a fragment of what is necessary, and we have to be bent on creating a healthy whole. If we could stand on Dead Horse Point five hundred years from now, that spectacular promontory above the vast canyon country of southern Utah would still astonish us with its expansive beauty. The only apparent difference might be that the slim, silver sheen far below, where the Green and the Colorado rivers now flow together, would be invisible, the water dried up and gone. Indeed, if our current practices continue, all our topsoil and all our clean water will be long gone. In too many places in our natural environment we have overgrazed our winter range, both literally and metaphorically. In our social world we have only a thin layer of topsoil—tolerance—for people and cultures that are different from our own. When times get hard, there is nothing left to nourish us, and we snarl over limited profits or limited resources. It seems clear that the work of spiritual regeneration has to begin now.

These intertwined aspects of life—the natural and the social, and the balance that comes only from our own personal interior—when taken together, seen as a single entity, become the habitat, an ecosystem, for a sustainable culture. When we have them in the right balance they will provide enough winter range to get us through the starving time, will make us strong enough to hold on till the coming spring.

For those who insist that there have to be three legs to our sustainability stool, we should at least rectify the names of those three legs, as Confucius suggests. Rather than Nature, society, and the economy, we can give them more accurate names: subsistence, sustainability, and spirituality. The reasons for this will become clearer as we continue, but if my notions about those three have any merit, they are appropriate for our stool because they are also inseparable.

For now, consider "Ask Me," a poem by William Stafford, part of a little series of poems about the Methow Valley in the North Cascades of eastern Washington, that has lingered for years in the workbasket of my thinking about sustainability.

> Sometime when the river is ice ask me
> mistakes I have made. Ask me whether
> what I have done is my life. Others
> have come in their slow way into
> my thought, and some have tried to help
> or to hurt: ask me what difference
> their strongest love or hate has made.
>
> I will listen to what you say.
> You and I can turn and look
> at the silent river and wait. We know
> the current is there, hidden; and there
> are comings and goings from miles away
> that hold the stillness exactly before us.
> What the river says, that is what I say.[6]

That last line seems to me the master key. We'll know we've got our environmental priorities; our society's social, political, and economic priorities; and our personal, spiritual priorities right when we have listened to the natural world carefully enough that we can say with Stafford, "What the river says, that is what I say." In *The Dream of the Earth*, Thomas Berry tells us, "The earth will solve its problems, and possibly our own, if we let the earth function in its own ways. We need only listen to what the earth is telling us" and then change our most interior life and practice.[7] Listening to the river is the way, the tao, that will both heal and empower our spiritual life, enabling us to transform our corner of the culture, moving it toward health for our environment and our society.

Another Stafford poem, "Where We Are," has always accompanied the first. If you know the territory, think of the Methow Valley, the high Cascades. If you don't know this particular territory, imagine your own high plains or timbered hillside above a stream. It will be entirely appropriate.

> Fog in the morning here
> will make some of the world far away
> and the near only a hint. But rain
> will feel its blind progress along the valley,
> tapping to convert one boulder at a time
> into a glistening fact. Daylight will love what came.
> Whatever fits will be welcome, whatever
> steps back in the fog will disappear
> and hardly exist. You hear the river
> saying a prayer for all that's gone.
>
> Far over the valley there is an island
> for everything left; and your own island
> will drift there too, unless we hold on,
> unless we tap like this: "Friend,
> are you there? Will you touch when
> you pass, like the rain?"[8]

Notice here that Stafford is talking about a special, though very common, kind of rain. Not a trash mover and gully washer, but a softer rain. This is no violent storm, confronting us with ruin, but the soaking rain we need after drought, a bringer and sustainer of life, the rain that heals.

Nevertheless we may still be lost: "Your own island / will drift . . . unless we hold on." Notice the shift in the pronouns here from "your" to "we." One glistening fact of our time is that we all need each other now—both the friend and the stranger—perhaps more than ever in the history of the world. The earth, and life as we know it, is peculiarly imperiled just now; we are all more vulnerable than we have ever been. The question then remains for all of us, "Friend, / are you there? Will you touch when / you pass, like the rain?" How we answer will determine whether we have a culture with a future.

Functional Cultures and Structural Cultures

GARY SNYDER OFFERED ME another way to think about learning native wisdom in an important observation he called "The Three Lineages":

> Mythopoetically speaking, there are three human lineages. One is the Children of Abraham —a group who all believe in a very convincing story, and one which traces their contemporary existence back to the somewhat quirky patriarch Abraham. These people of course are known in the world today as Jews, Muslims, or Christians.
>
> The next group is the Descendants of the Primates. These people are commonly found in universities, coffee shops, the upper levels of corporations, Hollywood, and the Democratic party. They are not necessarily comfortable with their relationship to primates—monkeys really do seem foolish sometimes—but they have no choice but to accept it, since it is seen as the only rational answer.
>
> The third group is the Sons and Daughters of the Bear Mother. This lineage dates back to the very far past, when the Bear Mother had children who were half human, and they in turn are the ancestors of the whole lineage. The Sons and Daughters of the Bear Mother are the most ancient and widespread of all lineages. All of North and South America, all of ancient Eurasia, and much of contemporary Africa. These are—you might say—the Pagans. These folks never had any problems accepting their kinship with animals—unlike the descendants of primates. The highly civilized current members of this line include Hindus, Confucianists, Daoists, Buddhists, Korean Shamanists (Korean national mythology actually says they are descended from bears), a few Turtle Islanders, and a lot of Japanese.

When I read this in an e-mail message, I thought, "Makes sense to me." I love the typical Snyder playfulness as well as the insight in it. It intrigues me because the "third lineage" gives me another way to think about what I observed, or thought I observed, during my early days in Alaska batting around in Eskimo, Indian, Aleut, and Tlingit villages. I've been struck again, especially in light of the environmental plight, but also considering the plight of our short-term culture and all its social and environmental relations, by the contrasts between those indigenous cultures and our own. It seems important to discover what we might learn from those much older cultures.

The matter now seems even more urgent than before. A decade or two ago, I still believed we had at least a couple more generations before our environmental destruction caught up with us, that maybe we had time to turn things around. Yet the recent revelations by climatologists about the potential of an ice age in the United Kingdom and northern Europe give me pause. And the worst-case projections, such as those reported to the Pentagon,[1] indicate that we may have only a couple decades, not generations, before climate change and our destructive tendencies catch up with us in more profound ways than our worst nightmares could predict.

When we think about sustainable cultures, we have models right at hand (as in Snyder's "Axe Handles"). Eskimo and Indian cultures in Alaska have been around about three times as long as Western civilization, and ten to a hundred times longer than any European empire that we study in history books. Now, given their imminent demise because of Nature's rebuttal to our demands upon it, or because of the machinations of transnational corporations that seek to rule the world, we nonnative peoples have to wonder, What do those indigenous peoples know that we don't? What might we learn from them?

A recent conversation with Wes Jackson at the Land Institute in Kansas provided some new language for this effort. Wes, who is working on developing a "perennial prairie" that could feed humans and other creatures and sustain itself over long, difficult periods, says, "We have to mimic the structure in order to be granted the function." I like the modesty of that. Not "we can get our hands on—or control—the function," but the function will "be granted" to us. I also was caught, especially, by the distinction between structure and function. Naturally, I have warped those two keywords in Wes's statement to my own ends here, using them in ways he may not have considered or may not approve. Neither he nor Gary Snyder is to be blamed for whatever foolishness I have created in what follows.

I want to distinguish two types of cultures: functional cultures and structural cultures. Both cultures have structures (complex kinship patterns, language systems, clans, moieties, ritual observances, taboos, courtesy patterns), but functional cultures are organized not only around them but around other elements of culture as well. Functional cultures include other cultural systems in their own: bees, birds, all the skitterish reptiles, moose, mice, and other mammals, the finned and flippered fish and other creatures of seas, rivers, and lakes that represent larger than human natural systems. Indeed it may be more accurate to say that members of a functional culture see their culture as incorporated in the larger cultures of a larger Nature.

However that may be, what follows are some sweeping generalizations without evidentiary support. Yet I do believe that one can observe these characteristics in traditional (functional) cultures such as I was privileged to experience in Montana and Alaska and have observed in my own Western (structural) culture, but I pose them as propositions, not proofs. They are concepts for consideration rather than conclusions.

Functional cultures are complex, balancing and holding together a huge web of information, intuition, ritual, spirituality, arts, humanities, and science to create a whole cultural system that operates within greater natural systems. None of these areas of insight are separate disciplines, and one may not be more important than another. Wisdom is the goal of learning. Citizens of structural cultures sometimes mislabel such knowledge "simple" or "unsophisticated" or "superstitious."

Structural cultures tend to put their faith in a harder, more structured knowledge called "science." ("Now what I want is Facts. Teach these boys and girls nothing but Facts," says the dunce inspecting the school in Dickens's *Hard Times*.) Structural cultures tend to suppress the disorder of intuition and to create order in the structure of their institutions, including their ritual or religious systems. That is, they are based on hierarchies, including codes. The codification provides the structure for both ethics and law, whether civil or religious. The hierarchical structure of the culture is replicated all along the way in social intercourse, in thought patterns, in the structure of the codes, either legal or ethical. In structural societies the arts and humanities, which tend to have a functional connection to wildness and an animating spirit (although with an emergent "form" that serves function), are less important than economics and science. They are often seen as frills, the first things cut when budgets get tight. Knowledge, which is less than wisdom in functional cultures, occupies an elevated status in structural cultures and becomes the goal of learning. The structural culture is thus both fragmented and simplified, and it lacks a comprehensive view of how the world really works. Citizens of structural cultures call this type of culture complex, sophisticated, and knowledgeable—without any apparent sense of irony.

The aim of education in a functional culture is wisdom that will ensure the survival of the people. The means of education are storytelling, dance, experience, imitation of elders, and careful observation of the natural world and of the human, social world that is included in it. All members of the culture participate in such observation, and universal education is the norm.

The aim of education in a structural culture is knowledge that will lead to personal advancement or advantage. The means are "facts," acquired via books or computers. Careful observation of the social and natural worlds is left to specialists. The human, social world is seen as distinct from the natural world. More education is available for some, less for others. The facts citizens of structural cultures want to learn are those that are useful in acquiring knowledge, which can be used to gain influence and power.

Functional cultures are organized according to consensus and appropriate scale, and their purpose is to function, to maintain balance and harmony in humans' relationships with one another and with the surrounding environment. The goal for members of a functional culture is generally to develop oneself as best one can, seeking to gain wisdom that will serve the community. "First we make sure everything in the system has enough, then we look after ourselves," its citizens say. Agreement is reached through dialogue and consensus. Functional cultures seek to utilize the gifts of the environment for food, shelter, equipment, and, in fruitful times, luxuries. They try systematically to avoid altering the structure of the environment, especially its balance and harmony.

Structural cultures are organized according to hierarchy, a structure that the culture assumes works best. Its purpose is growth: increased income, territory, power, and property—including other people, grain in the bin, the gewgaws that come from commerce. The goal for citizens of a structural culture is generally to develop certain skills that serve the self first, so that one can find work to support oneself and one's immediate family, helping it acquire the things that reflect success. If there is enough left over, out of its abundance citizens in a structural culture may share with others. Agreement is reached not through dialogue but by vote, or compelled by imperial edict or by legislation. Structural cultures tend to alter the function of the environment for convenience, ease, or profit.

Functional cultures have members (in many tribes within the United States, now "enrolled members") who belong, all of whom are enfranchised to participate in public life. These are members in the sense of family members, recognized as such by the community.

Structural cultures have citizens who may or may not participate in public life, and some have no role in civic life because they are disenfranchised by poverty or lack of education, race or sexual orientation. Those who live in a structural culture become citizens by various kinds of legal certification rather than recognition.

The functional culture seeks to live within natural systems so that they provide sustenance and physical safety: food, clothing, shelter, and protection from wild animals and other cultures. Functional cultures also seek spiritual safety: safety from giving offense to the spirits, animals, and plants of a place that might upset members' own inner spirit and undermine or destroy their physical and spiritual relationship to the world. Functional cultures know that if one's spiritual relationship to Nature is undermined, one's physical connections are jeopardized. Appeasing the spirits is an important cultural goal. It may be simple fear—the fear of spiritual ruptures—that drives functional cultures to seek to appease the spirits.

Structural cultures also seek safety. Safety from the ravages of the natural world is to be wrested from a hostile environment by knowledge rather than wisdom. Citizens seek to protect their wealth and their persons. Spiritual safety is won by obedience to hierarchical institutions and encoded beliefs. Structural cultures believe that humans have only physical, not spiritual, connections to Nature. Structural cultures seek to appease their institutions, the very ones that alter the structure of the environment for the convenience, ease, or profit of their institutions' executives, the upper levels of the hierarchy. It may be simple fear that drives structural cultures to appease and thus protect their institutions.

Both cultures seek power. Functional cultures seek the power inherent in good relationships. They try to find their place in the system in which they live. They want to stand in right relationship not only to the human community but to all the creatures in the natural system. They are horizontally integrated: their own structures are part of the larger system and the other natural cultures that surround them, everything connected and reaching out to affect and include everything else. Some scientists and religionists today talk of "horizontal transcendence" rather than vertical transcendence, to underscore that spiritual connections are contiguous with environmental and social connections.

Structural cultures seek the power inherent in dominion. They try to create a place in the system that will give them what they want from other cultures and from Nature, both of which are seen as resources. They want to maintain a more powerful stance than other human societies and to be vertically integrated in the highest positions in their own system, regardless of what that may mean to other cultures, human or otherwise, around them. Vertical transcendence is the order structural cultures see in the natural world, with the human species at the top. That vertical transcendence matches the view of humans' spiritual relationship to the godhead that exists "up there," with humans just a little lower than the intervening angels, above all other creatures.

Functional cultures are organic, natural, quadriform: four seasons, four directions, four colors . . . They are less time constrained and more seasonally conscious. Time is not so much incremental or digital as flowing. It does not move in a linear fashion but may curve around. The past reoccurs in the present; elders who have gone before may be very present.

Structural cultures are engineered and are generally trinitarian. Christianity's trinity is but one example. It seemed only natural for Freud to find an id, an ego, and a superego. I have seen Anglo teachers read an Eskimo student's paper in which an idea is examined from four perspectives and automatically redline the fourth as redundant. In structural cultures time is minute conscious, its passage now marked in digital clicks on the wrist. Time moves in a linear fashion; the past is past. Even when elders or old friends, now deceased, return, they are not present but exist only in memory.

In functional cultures everything is connected, and the questions for study are, What is the function of this? What relationships are we seeing here? Functional cultures seek to discover connections in the nature of things and understand that "Ground Squirrel is grandfather to Bear," as Tlingit elder Austin Hammond said one day when I was visiting at his place.

Structural cultures study structures, whether human or otherwise; they isolate and disassemble them. The questions for study are, How is this made? What is it made of? What are its components? Rather than study the relationships within and between things, structural cultures take them apart, searching for smallest units. Until ecology came along, structural cultures pretty much ignored exploring relationship or function in favor of analyzing structure. Even now structural cultures are puzzled at the notion that Ground Squirrel is grandfather to Bear. An example of this attitude might be the approach to something like Ebola. What structural cultures seek to do is isolate it and destroy it because it harms humans. Structural cultures do this without asking what its role in the larger Nature might be or whether it could be essential in the lives of other creatures that may be essential to our own.

In functional cultures everything that has life has a spirit; thus all the gods have at least one characteristic in common: life, in both its physical and spiritual aspects. Since all things are connected, everything has to be treated with respect, even reverence. The gods are not transcendent but local. They are the gods of the place. The gods of the desert are different from the gods of humid places where water is abundant. If the people move, they discover and respect the gods of the new place. This sustains a concern for not upsetting relationships, for keeping the taboos, maintaining certain rituals, and paying attention to intuition and even dreams. The gods are not necessarily eternal. Therefore they need not only reverence but care. They nourish the people, and the people nourish them. Functional cultures honor the spirits in ritual and in daily behavior, which takes place where the spirits reside.

In structural cultures humans are the most important creatures, and everything else, because it is lesser, is less deserving of respect or reverence. Even lesser humans, those further down in the hierarchal structure, deserve less respect. The gods, or God, is transcendent. They too have a hierarchy in ascending order from saints to angels to the godhead. Faith (belief) is torn between thinking the gods are alive and thinking they are fantasy. They are objects of veneration on the one hand and superstition on the other. Because spiritual realities are eternal and separate from earthly realities in structural cultures, structural divinities do not need humans' care but require our faith. Structural cultures honor their gods in ritual and in their institutions and tend to ignore them in daily behavior. Daily behavior takes place in different institutions from those in which the gods reside.

In functional cultures the purpose is to find the people's place, to fit into natural relationship systems in a way that maintains health for all.

Of course structure exists in functional cultures—though in a less stilted form—as in a conversation in a village about a serious matter, or in an elaborate courtesy pattern that inhibits conflict. In functional cultures such structures unfold naturally and remain servants of function.

In structural cultures citizens impose structure on systems not yet understood, believing that they have the right to create what appears to be order out of what appears to be chaos. Instead of having a function that serves the health of all, what is natural or organic is used, "developed," and pressed into service of the hierarchy. In contemporary U.S. structural culture we ape the worst years of the caste system in India. The poorest get nothing, few of our Brahmins care, and our major economic institutions, including government, design themselves to see that the lowest poor in the hierarchy get even less, while the top gets more.

Control in functional cultures remains local and resides in the community. The "chief" often has little authority, and the shaman may scare a particular family into bringing gifts or doing something he says and may cause wariness or fear around him, but for the most part the entire local community participates in controlling the local community. No one person or class has a civil authority to compel others to a particular course. Control is not the main function of the social structure; balance and harmony are.

In structural cultures control is a primary goal of the structure. Control resides in the upper echelons of the hierarchy that dictate to the rest of the community. In the political arena, that control may rest in the military or in an executive, legislative, or judicial entity. The economic sphere may be controlled by a CEO or by a transnational corporation with its own structural hierarchy. Those controlling the local community are rarely residents of that community; a great deal of community control resides elsewhere, often far away. This is the fundamental nature of colonialism, and colonialism is a primary characteristic not only of the governments of structural cultures but also of their major religions. In structural cultures evangelism and conversion of the heathen become euphemisms for colonialism and economic exploitation.

Functional cultures trust personal experience and observation. They are highly adaptive; they change according to observations on the ground and their own experience, especially as other observers share and refine their information. Even long-held traditions, the "old ways," may shift under such scrutiny and dialogue. Because there is no vested civil authority, each new community or tribal engagement must be thought through by the community. The question becomes, Who speaks with the greatest wisdom in this situation? In other words, who is the real authority here? That person is not always the same from event to event. The last successful war leader may not be the one whose advice will lead the people to the best food sources. The best hunter may not be the best at restoring balance and harmony in the community. Thus community members are not easily led; they listen carefully to one another, and they tend to talk at length. Discussions may continue for days. Oratory and rhetoric become vehicles for examining issues. The aim is to think things through and then to share careful thought so that everyone moves toward wisdom.

Structural cultures trust their structures more than their own experience. Because they trust the structure, that is, their institutions and institutional authorities, citizens are easily managed and manipulated by "leaders"—even when those institutions and authorities are working against the community's best interests. Civil authority seeks to have its way, and institutions rather than individuals become the vehicles for examining issues. Oratory and rhetoric become advertising, means of selling the community on the direction that its hierarchy wants to pursue. The purpose of oratory, political or otherwise, is not to explore and think through but to persuade, whether the claims are true or false. The question becomes not who has the wisdom but who has the most persuasive public relations firm. Citizens in structural cultures are sometimes persuaded by a perception of which side will win. Because they trust more in the structure than in personal experience and thought, structural cultures first become rigid (thoughtless, without self-examination), then become dysfunctional, then topple. Sometimes those in power begin to believe their own sales pitches and thus become as ill informed as their poorest citizens. Then they too are ignorant, rigid, dysfunctional—and they topple.

In functional cultures persons take precedence over institutions; community takes precedence over individuals.

In structural cultures persons are secondary and succumb to institutions; community is secondary and succumbs to commerce or corporations or rulers of other institutions.

Functional cultures understand the difference between governance on the one hand and economics and commerce on the other. Economics and commerce, though important, are not seen as the driving forces of all human endeavor, and leadership in trade is not seen as an asset in political leadership. Education and politics, for example, are understood to be fundamentally different from business.

Structural cultures, soon or late, come to see everything as a business. Education and politics are seen as commercial enterprises. Structural cultures elect business leaders (even failed business leaders) as governors and puzzle over why political rhetoric degenerates into the rhetoric of advertising. The traditional rhetoric of commerce becomes the lingua franca of government; language spins. When the spin becomes too great, language falls out of control and loses its relationship to reality.

Functional cultures assume that all things have essential needs that must be met to keep the whole natural system, including the human society, in balance and harmony. They recognize that natural systems support the people and try not to offend or alienate other "people"— four-leggeds, bird people, rivers and lakes, the living soil. Other such "people" are considered parts of the human family.

Structural cultures assume that human needs are primary. Impoverishing natural systems does not worry structural cultures much, and they deplete them without understanding or checking what their loss might mean, not just to humans but to the natural systems of which we are an organic part. Other creatures are truly "other," are often seen as repulsive, and are not considered part of the human family.

Functional cultures are long term, what we would now call sustainable. They pay attention to everything around them, taking into account all creatures and landforms, both at home and afield. Thus, trusting their experience, they are able to maintain their function in their ecological niche and allow other creatures to continue to function as well. So they often last for extraordinarily long periods (Eskimos and Aleuts, for example, at least fourteen thousand years thus far).

Structural cultures are comparatively short term, perhaps five thousand years thus far. Because structural cultures tend to take human things into account but ignore much of the rest, except for their use as resources, they do not see the connections between humans and everything else in the complex environment that surrounds us. Structural cultures assume that Nature really is separate from humans, and therefore any impact we have on it is beyond any negative effect that could spill into our lives. As a result structural cultures can unravel whole environmental systems without knowing till it is too late (by effecting climate change, for example). When that happens, the structure cannot keep its balance, and the natural world topples. In the social world, for another example, a growing gap between rich and poor is the first mark of the apogee of the arc of an empire's structural lifespan. When the gap grows too wide, the center cannot hold, and inevitably, things fall apart; the structure collapses in violence and chaos.

When things get out of whack and illness strikes, the healing that functional cultures seek to employ is aimed at restoring balance and harmony in relationships. The shaman's assumption appears to be that if the people make their relationships with Nature and the tribe right, individuals will be all right. So the purpose of the Navajo night chant, for example, as I understand it, is precisely to restore that balance and harmony, to learn to walk again "in Beauty," reincorporating and thus incarnating one of the essential components of sustainability.

Structural cultures aim at healing the self first. Psychiatrists assume that if individuals get themselves together, their relationships with other humans will automatically improve. This psychiatry does not suggest that humans might be out of touch with, out of balance and harmony with, the rest of our environment beyond the humans in our circle, or that we might need to restore our relationships in both spheres to restore health to our souls.

Both functional cultures and structural cultures may change over time, taking on characteristics of the other. Functional cultures may, under the onslaught of dominant structural cultures, become more structural. This may be a temporary survival technique, or it may actually reflect the adoption of the structural culture's value system. When the latter happens, the functional culture is lost. A subsequent generation may seek to reclaim the old ways, but that process is very difficult, especially if the language has been lost in the transformation from functional to structural. The usual end of such an effort is to adopt the trappings of old ways (dance, music, a few stories) while maintaining the new structural core. The old functional culture then becomes a thin veneer, a gloss of old ways without substance, appearance rather than reality, ultimately an imitation of an imitation.

Structural cultures also change. Western cultures were once functional, that is, animistic. The earliest Western philosophers, particularly the pre-Socratics, saw the earth as a single living organism. Philosophy began as "natural philosophy," seeking to understand the whole natural world around the culture. But it became lopsided instead of fully rounded. It began to rotate like a cam rather than a circle, and it was impossible for it to maintain balance. Functional reverence, care, and wisdom became subordinate to structured power, economics, and knowledge. Control of relationships for the benefit of a few humans, rather than maintenance of relationships for the benefit of all life, became the goal. Unwittingly, functional cultures that became structural cultures thus put limits upon their survival. Much in structural cultures is veneer; packaging becomes more important than product.

Nature is inclusive rather than exclusive, integrated rather than fragmented, a flowing, flexing, evolving whole rather than an assembly of constituent parts. Functional cultures mimic natural cultures, trying to alter their own behavior so it fits into Nature's behavior. They have a function in the whole scheme of Nature, just as everything in Nature has a function of its own, all of it in relationship to everything else.

Structural cultures, in contrast, are disruptive and exclusive, altering Nature to their own ends rather than altering their own behavior to fit Nature's. They play an artificial, adopted role, but they have no useful, authentic function in the whole scheme of Nature.

If this comparative system has any merit, the third lineage Gary Snyder describes above is a functional culture; the first two are structural. There are, thank goodness, still functional descendents of the Bear Mother. Perhaps we will yet learn the wisdom of their ways.

Subsistence

Exploring Subsistence

SUSTAINABLE CULTURES ARE TO subsistence cultures as squares are to rectangles. That is, all sustainable cultures are subsistence cultures, but not all subsistence cultures are sustainable. Here I explore the nature of subsistence and look at what we might learn about our own sustainability from that exploration. Of course I do not think we should—in some romantic dream fog—try to return to the old ways of hunting and gathering. Nevertheless, I want to explore a tribal definition of subsistence by looking closely at contemporary traditional cultures with which I am somewhat familiar. From our place in a subsistence culture, we can see what it takes to move toward becoming a sustainable culture as well. In some observations of traditional subsistence, we will find parallels with our mainstream American culture, for we are as dependent on the land as any traditional people ever were. Our hunting may not be for animals, but we are hunters and gatherers still. We may not see ourselves as such, but our frantic hunt for the last barrel of oil is not a metaphor but is very real hunting and very real gathering. Oil is our big game, and the dangers in our hunt for it and the consequences of our failure to find it are as real as those of a caribou or whale hunt are for Eskimo people. Without oil we are as apt to be hungry as an Inupiat village without whales and seals. Our culture is a subsistence culture unaware of itself.

If subsistence is our culture, and if we are to live it fruitfully and use it as a model to create a sustainable culture, we may turn again to those traditional cultures that have much to teach us. The ones I know best are those of Alaska's native peoples, though this does not mean I understand them. Perhaps no one outside the culture can make that claim, and some inside the culture cannot either. That is only natural, and as true of modern American culture as any other. Stop any person on the street and ask him or her the difference between Plato and Aristotle, and few could answer.

What I am looking at here, trying still to understand, are the intriguing glimpses of the various cultures I was privileged to witness over a quarter of a century. The subsistence life of village people has changed over the years, as every culture changes. Subsistence now requires some cash, just as living in mainstream American culture requires more money than it used to, for items we used to do without: refrigerators, television, autos, health insurance . . . Many village people now live a hunter-gatherer life mixed with a part-time or full-time job. Yet much of the food of people in small villages on the lower Kuskokwim River or along Bristol Bay in Alaska, for instance, is still derived from land or sea by the work of their hands, the risk of their lives, and their incomparable knowledge of the region. Many still participate in rituals and ceremonies, dances and songs that reveal a life embedded in the land and share a profound and complex belief system that includes animals and that is central to their daily lives. That worldview is the primary aspect of a subsistence life; technology and cash are only adjuncts, the tools of hunting and gathering or the medium of exchange. Much of the cash they need to continue their subsistence life in modern times comes from the harvests of land or sea, or the handicrafts and art that are made from raw materials garnered from the hunt or harvest. Often that income is supplemented by work for a native corporation, a local health agency, or the school district. That they work for cash to run the outboard or buy shells for the rifle does not mean that they are no longer a subsistence people. Cash income from such work is simply incorporated into the subsistence life. Cash and subsistence are not contradictory terms, in part because cash is limited to the economy; subsistence transcends the economy.

Sometimes Anglo observers of the subsistence life remark on

changes in technology and point to the use of rifles, skiffs, and kickers to indicate that the old ways are gone and that the subsistence life Alaskan Eskimos, Indians, and Aleuts have lived for thousands of years is over. Since it is not technology that makes a subsistence life, however, such a notion is misleading. For indigenous peoples subsistence means direct, personal engagement with land and sea and the recognition of humans' dependence on the land for energy, nourishment, tools, and household goods. Our American culture no longer demands such personal engagement, but ultimately all subsistence depends not on technology, not even on fish and game, livestock or agriculture, and certainly not on transnational corporations or oil, but on persistence in maintaining a certain worldview. The difference between a harpoon and a gun is irrelevant to the necessity of hunting seals, as is the use of a boat covered with walrus hide instead of aluminum. Aluminum skiffs and high-powered outboard engines have been adopted by these adaptable people for many of the same reasons that we adopted automobiles: ease, extension of range, speed of transport, sheer pleasure of power. When we took up those purring motors, we did not realize that they would power us to a dead end. It is life on the land and the view of the world, not the technology, that create subsistence and sustain a culture. Since our culture no longer maintains such ties to the land, what it might adapt to its advantage is the worldview.

Non-Indians have made much of the differences between American Indian perceptions of the land and their own. What is often overlooked is a fundamental similarity: that we all live in a subsistence relationship to the land, that every economy, regardless of cash, credit, stocks, or other media for the exchange of goods and values, is but one part of a subsistence life based upon what the land has to offer for energy, raw materials, and food. If ours is not a subsistence culture, why do we fret so over oil supplies, timber harvests, commercial fishing, the number of cattle on our public lands, the depletion of our aquifers, the poisons in the air we breathe, and the toxins in the water we drink and the food we eat? Clearly we are dependent on the land, and our own cultural existence transcends our economy. Ultimately our culture's life depends upon a healthy landscape as surely as any traditional people's ever did. It also depends on our worldview, our sense of place and purpose, of how we might best fit into an ongoing life within the constraints of a Nature we have thought of as either a lovely landscape that restores our souls or a lucrative landscape that is ours to exploit for personal or corporate gain.

Defining Subsistence

One of the problems in our discussions of subsistence is much like a problem we have with sustainability: it is one of definition. It is easy to see what subsistence is not; it is not, despite anthropology's nomenclature, an economic system. I sat in on a conference in Juneau a few years ago where village people and nonnative hunting guides came together to talk about subsistence. The nonnative participants referred to subsistence primarily as "hunting, fishing, and berry picking," a phrase that came up repeatedly. Their definition was basically an economic one. Native participants, on the other hand, consistently talked of subsistence as "a way of life." Others insist that subsistence is not a merely a way of life: "It *is* our life." That life includes far more than hunting, fishing, and berry picking or even the most intimate knowledge of land and sea and their animals and plants. We nonnatives are also likely to view subsistence as a level of existence, as in "They were living at a mere subsistence level." We see that level as lower than our own, less sophisticated, more primitive, even "superstitious." Native people see subsistence as a base upon which an entire culture establishes its identity: philosophy, ethics, religious belief and practice, art, ritual, ceremony and celebration, law, the development and adaptation of a variety of technologies, and an educational system that will ensure the survival of the people. All fall within the realm of subsistence. The gulf between the native and nonnative definitions is profound.

Nor is subsistence just a conservation-based ecological approach, allowing people to survive on the land. That resembles the notion of stewardship, and therefore we applaud it. But "stewardship" is a limp word to equate with subsistence. Subsistence, as conceptualized by

Yup'ik people, is immensely more complex and more insightful than acting as sty wardens of the world. As anthropologist Ann Fienup-Riordan
notes, "Given this cultural framework, it is possible but altogether inappropriate to reduce subsistence activities to mere survival techniques
and their significance to the conquest of calories. Their pursuit is not
simply a means to an end, but an end in itself." Stewardship puts the
power in people. But, Fienup-Riordan says, "The real power is not in
people, but in the continuing relationship between humans and the
natural world on which they depend."[1]

Subsistence is not only hunting and fishing but the life that gives
meaning to hunting and fishing. History and metaphysics are also involved. Birth and death, and a constant rebirth and dying, are central
to that continuing relationship and reciprocity between humans and
the natural world that Fienup-Riordan observes. That relationship, often trivialized by Americans' faith in industrial systems, is maintained
by practices we scorn as primitive. Yup'ik children, for example, are
named for deceased persons, and the souls of the deceased are thus
alive to those still living. The souls of seals or salmon, of moose or
ground squirrels, when treated properly, are able to return another season. Thus, in a sense, the relationship has been maintained by the same
souls, the same humans and seals or salmon, who have been interacting
all along. "The birth of a baby is the rebirth of a member of its grandpa-
rental generation. The death of the seal means life to the village." Thus,
says Fienup-Riordan, "The same people and the same seals have been
on this earth from the beginning, continually cycling and recycling
through life and death."[2] I'm sure I don't have all this exactly right,
which is why I keep trying to think about it. But perhaps it is right
enough to indicate a connection between Nature and humans that is
both more profound and more insightful than we suppose when we
disparage subsistence as a simple culture's economic system. Subsistence ramifies through every aspect of the culture and knits the culture
whole. As long as indigenous cultures and other cultures continue to
use such differing definitions of subsistence, there will be no communication about environmental or any other issues.

The narrow concept of subsistence as an economic system comes primarily from anthropology. It does not come from economics, for economists are not out in the field studying other cultures. Nor does it come from history or other social and behavioral sciences, though these disciplines have adopted anthropology's definition. Anthropology also took the lead in developing the taxonomies by which we divide and subdivide the lives and cultures of other people for study. This seems natural to Anglos, whose cultures have been fragmented for centuries. But these taxonomies are "outsider" names; they do not come from within the culture, and they do not represent definitions that traditional peoples would use. They are names for non-Indians' preconceptions of what we expect to find. Our names for Eskimo culture represent fragments that were once parts of a whole for us and are still parts of a whole for indigenous cultures. Continued use of anthropologists' definition of subsistence perpetuates a misperception as if it were true. Both the misperception and the definition are destructive of traditional subsistence life and pernicious in the extreme, as anthropologists are often enlisted in the Division of Subsistence of the Alaska Department of Fish and Game to study subsistence practice and offer advice on policy. Anthropologists in the division spent considerable time helping folks in several villages record the numbers of caribou, moose, ptarmigan, ducks, geese, various species of berries, salmon, herring, and parky squirrels that were used over the course of a year. These clearly economic measurements reinforced the notion among nonnatives that subsistence is an economic system.

A further complication in thinking of subsistence as an economic system is that such systems are usually associated in nonnative minds with so-called primitive cultures. This may stem, again, in part from their association with anthropology, which has been seen for years as a discipline that studies cultures that are "primitive," perhaps even "savage," or at least "exotic." We often say of such peoples that they must move into a cash economy, which, in our view, is more sophisticated and complex—and like our own, of course. We want that for native people now, precisely at a time when our own culture is moving to a credit or even cashless society where transactions are increasingly made by computer-generated electronic transfers.

If subsistence is a way of life, or if it is indeed life, it is dependent upon the way people live, not on the availability of fish and game or certain vegetation. In the old days when game was gone, subsistence life did not stop; even when the game did not come all winter, when the rituals and prayers and ceremonies all failed and people starved, the survivors persisted in a subsistence life. There was no other choice for them, and, in fact, there is none for us.

Subsistence, then, is not an economic system, nor a matter of numbers of fish and game harvested in the course of a season; it is a matter of worldview. The white man's preoccupation with economics and determination to quantify everything are irrelevant to the life under discussion. It is possible that we could have, in the future, the largest populations of moose and bear, caribou and salmon, that Alaska has ever known. Sound wildlife management might bring us that highly unlikely outcome. But we can have abundant game and have no subsistence culture left. If that happens, traditional culture will have slipped its moorings and lost its way, become part of mainstream American culture in a massive act of assimilation, joining an exploitation culture that has forgotten its roots in the sacredness of Nature. What is critical for all of us is to know ourselves—regardless of the culture we belong to—to be subsistence people, to acknowledge the subsistence character of the culture, and to practice a life that protects both the land and the culture through its beliefs, rituals, songs, and stories. We must be willing to take time for the rituals even when hunger gnaws us and dogs our families. Western culture has deliberately struggled to forget the rituals to set itself loose to pursue every avenue to take from the land anything that smacks of profit, without due consideration for the land itself. Yet depleting resources is like depleting a bank account; sooner or later we go broke and belly-up.

The External Threats

Both anthropologists and educators, in the name of offering understanding, sympathy, and help, have been in the forefront of the effort to eliminate subsistence. In a tragic way anthropologists and schoolteachers have a common record. Both have often meant well, and tried to stand against the standard-brand racist institutions, agencies, and personnel, all of whom (me as teacher and bureaucrat included) have been the bad guys, because no matter how well meaning or astute or careful, we just don't know what we are doing when we tinker with other cultures. Social engineering always has unanticipated outcomes, and the most important results are almost always negative. Perhaps that stems from the arrogance that lies behind every social engineering effort: the idea that someone or some small group, often outsiders, knows so much better than others the life that they ought to live. The result of our American engineering effort has been that the academy has laid siege to traditional societies, and the knowledge of them gained and disseminated by anthropology and other disciplines has often been used against them.

The public schools also have been determined agents for the erosion and eventual loss of indigenous cultures. An Alaskan superintendent of schools once insisted to me that his purpose was "to wipe out the last vestiges of Eskimo culture as fast as I can." For years the goal of schools for American Indians was assimilation and ethnocide. One of every culture's most profound characteristics is language. For years in Alaska we insisted that village children forget their own language and adopt English. The teachers punished any use of the local language on the playground or in class, sometimes severely. Washing a student's mouth out with lye soap was a standard punishment, and not the worst. Nothing diminishes a culture more quickly than the oppression of its language.

The Internal Threat

Despite the external threats that have been hurled at indigenous populations in the Americas for hundreds of years, the threat I fear most is no longer an external one. The gravest threat to traditional culture now is an internal deterioration of subsistence: the withering away of subsistence values from within the culture; the loss of subsistence integrity, not forced by nonnatives but coming from the loss of knowledge and understanding inside the culture. That is the way that Western cultures lost their own sense of subsistence. This was not something that somebody else did to us; we did it to ourselves. There is a kind of romantic notion that if we can control the outside forces that have an impact on a culture, everything will be OK. But most often our most serious cultural losses come from inside the culture. We lose track of our own internal resources. Now my fear is that the stories, the worldview, the methods, and the magic that make subsistence possible will be forgotten. How do we maintain them? Since we all live in subsistence cultures whether we recognize it or not, this is an important issue for all of us.

If a traditional culture lets go of the life that gives meaning to subsistence activities, then it has also let go of its own definition of subsistence and reduced it to an economic system. In Alaska laws have been passed in recent years that gave hope to native peoples that the land could be saved and that subsistence would be saved thereby. The Alaska Native Claims Settlement Act (ANCSA) of 1971 and the state laws regarding subsistence priority for village residents are examples. But protections for subsistence culture cannot come from ANCSA, village schools, Indian Reorganization Act councils, or any other laws or institutions external to the culture. Protection can come only from living the subsistence life itself, from maintaining the subsistence worldview, from telling the real stories of subsistence. The law can only slow the white onslaught on traditional culture. Only the culture can protect the culture. The culture protects itself only by continuing—by deliberately choosing, each day, to continue—the training of the young in the subsistence values of the culture and by doing that from within the culture, despite what the schools or any other alien institutions of other cultures say or teach.

In part, Eskimo and Indian people have kept subsistence alive, and sustained their cultures, by learning from their elders. That learning has been primarily through stories, music and dance, and observation and practice in the field and at home. But it is not entirely a matter of learning what the elders have to teach, and it is not entirely a matter of stories and music and dance. That learning is absolutely essential, but real learning must also come from within, from personal experience, from seeing for oneself. Part of the genius of traditional educational systems is that they make provisions not only for texts (stories and songs) but for personal experience. Maybe we have to learn everything twice: once from others, and again in our own experience. This is important so that when, as elders, we pass our learning on, we speak not only from our learning but from the conviction that arises from an authentic life of personal engagement and experience. Youth then not only need to hear what the elders have to say in their stories but must go out on the land and develop their own close observation of the natural world, their participation in the lives of animals, in the prayers and practices that make the animals return. Much of that observation of elders working on the land, living the life, is a matter of learning how to pay attention to details and to relationships, of finding the connections between humans' behavior and the behavior of animals and plants. If that does not happen, then essential knowledge—the knowledge that holds the world together—is lost.

At an Alaska Federation of Natives evening of dancing, drumming, and singing, everyone was excited to see the kids up and dancing. Some village dance groups were composed entirely of junior high and high school kids. Folks were elated that they were learning the old songs and dances, yet as I sat there watching, I got more and more depressed. The kids were all lined up like a chorus line, everyone making exactly the same movements, like the Radio City Rockettes, singing songs and dancing dances they learned in school, taught by their elders. In the village community everyone, of every age, participated; in school it is only the kids and one or two elders employed as teachers' aides.

That, it seems to me, is the problem. In the old days a man prepared himself and his equipment, went hunting, came back, invented his song and dance, and performed for the whole village what was an integral part of a cultural whole, not something he learned from someone else. A woman went berry picking and met a bear, and that evening she turned the experience into a dance, and the dance was as integral to the subsistence of the people as the food. Those kids at the AFN convention had been taught, not in the community but in school, a fragment of what is too broad ranging and integrated to be taught. It can be learned in a community but not taught in school. It can be learned only in a system that is *of* the culture, not an institution imposed by another, frequently antagonistic, culture.

Anglos can be taught the elements of our cultures (art, music, history, language, etc.) a piece at a time because they have been fragmented for a long time. We can learn about our cultures in our schools because they are our own cultures' institutions. But they are not Yup'ik, Crow, Inupiat, or Navajo institutions.

The traditional hunter grew up in an ecosystem, a natural and social context, in which he could move comfortably. He learned about that system of relationships both from stories and by direct observation enabled by personal experience, participating in the process with an elder. Within the village culture, what the hunter learned was the whole process; what the school teaches is the song. So schools try to teach what can't be taught, and educators think they do it successfully, in part, because even the elders think it is wonderful. At the same time, schools deprive kids of an opportunity to learn the integrated whole because they have to be in school, which is where, in the dominant American culture, education takes place. So they don't hunt, not over long, contiguous days and nights out in fall or winter's dwindling light. Life on the land, engaged in subsistence activities, is the source for understanding the game and the landscape, and of both the song and the dance, which are integrated not only with the hunt but the preparation for it and the return in triumph or failure: the carving, the prayers, the making of tools, the planning and mapping, the hunting itself, the rituals of acceptance of the gift of animal life, the return, and the community celebration—out of all of which come the song and the dance. No, that's wrong: the song does not come out of all that; it remains part of all that. No single element can be pulled out without unraveling the whole.

As long as nonnatives persist in believing subsistence is an economic system, native peoples only play into their hands and reinforce the misunderstanding when they base their arguments for subsistence on the need for fish and game. For years anthropologists working for Alaska's Division of Subsistence furthered the mistake by counting fish and game taken during the year by native villages. To say "We need 1,900 pounds of birds and game to keep us for the year" is an economic argument that leads to the nonnative numbers game, balancing one economic cost against another. It is also an anthropologic activity that scholars in an alien culture know how to pursue, so they do it out of a desire to help. What it does, however, is reinforce the impression among Anglos, especially anthropologists, that subsistence is an economy, and it deepens the difference between our ideas of what subsistence means. It also works against traditional cooperative systems, for "quantities are competitive; qualities are complementary."[3] That is also one difference between an economic system and a people's life together. If song, dance, and subsistence can be broken out and described separately, then maybe they can be taught, but not as integers, not as integral to the whole, not with integrity.

More important, and more frightening, if they can be broken out, they can also be replaced. If subsistence is only a matter of economics—of hunting, fishing, and berry picking—then one can substitute army beef for buffalo (as happened in the lower forty-eight), beef for caribou or moose, or Spam for salmon, a shift one congressman touring Alaska noted was "easy to make." Even if the nutritional value is equivalent (though it rarely is), the culture that ramifies from subsistence and through which subsistence permeates is gone. Such a shift also makes village people dependent on another culture's food, the sources of which are miles away—a dependence on nonnatives that our larger American culture says it deplores but always encourages and that its laws often make inevitable. The sum of the parts of a fragmented culture always add up to less than the whole. An integrated culture adds up to more, a whole so large that the parts are lost or submerged in it, undiscerned by members of the culture.

Further, the bits and pieces cannot, perhaps ought not, be taught unless the whole can be taught, and it seems clear that the latter is far beyond our schools' willingness and capacity. In school, dance becomes dance only—a replaceable fragment. If a song is only a song, it can be taught in school, but if a song is the culmination of making tools, of hunting, of cold and endurance, of disappointment or triumph, of going out and returning, of feeding the people, and of expressing what all feel—if it is a community's life—then it cannot be taught except in that context. It surely cannot be taught in an institution that remains fundamentally alien, and alienating, to the community. To hold a class "about" it in school, even to have the elders teach the class, is to perpetrate a falsehood (we can teach what can't be taught) and to create false hopes (we can count on schools to keep the culture alive). To admit that the song is a fragment, to consent to that, is to condemn it forever to be a fragment, cut off from the rest of the culture and replaceable.

No matter how much the elders do through schools, the second leg of knowledge, which comes from the experience of life within Nature, is still missing for many Alaska native children. Subsistence learned in school is only a matter of hearsay, not experience. As such it is poor fodder. The analogy for education goes like this: school is to hearsay as local culture is to experience. The best learning about subsistence takes place outside school, in the field, youth and elders together, the elders helping the young ones to understand and interpret what they are seeing. In the field—not just hunting and fishing, important as they are, but being in the field, away from home and village—elders demonstrate to youth a worldview by living it rather than telling it. Back home later, they share it in stories and songs in a larger community context where every activity makes more sense than it does in a classroom. Shinto farmers in Japan share a similar notion. There, although traditional learning about the land is important, says the Shinto philosopher Munetada Kurozumi, "if you really want to be a good scholar for the benefit of the world by comprehending the real things, you must read the real book in the form of the 'living world.'"[4]

Replacing the Irreplaceable

When the culture is taken apart for observation or analysis by anthropologists or educators, the culture at that instant becomes in some ways a Humpty Dumpty fallen, never to be made whole again. Once the anthropologist says "Tell me about your dances," the members of the culture see dance in a new and isolating spotlight. It is not possible to return from the alien vision to see the whole again. Elders applaud the children's learning to dance in school because they don't think of it as a reproduction of a reproduction, a stark marker of the death of their way of life. They applaud because they have learned from nonnatives that this is "Indian art," or "Eskimo dance," activities once so unselfconscious as to be undifferentiated parts of a culture with integrity.

When village people sense the loss of wholeness, of integrity, they can only look back, never at, or toward, the whole as it exists now, or see it as it will or might exist in the future. If they could look at it as nonnatives look at their culture, they would be outside it, not in it—another feature of the Cartesian worldview and a sure sign that the traditional culture has already been diminished, the traditional worldview dislocated. What has shifted in our Cartesian view of the world is not our participation in Nature but merely our point of view. Our vantage point has become detached, that of a spectator rather than a participant. This is part of the problem as well: what we grope for, regardless of culture, is an integrated view of the world that sustains us.

Once the various elements of culture are separated, some parts seem to be more important than others—especially to an outside observer. They can be replaced by something like them (food for food), or they can be dropped entirely. We saw how this worked during the 1980s, when the imagined or invented necessity for defense took precedence over the arts. That precedence has expanded enormously in this new century. If art is separate from the rest of the culture, and if defense is the priority, then defense tends to become the entire cultural core, and art is seen as impeding the flow of scarce resources to defense. Art thus becomes not just a fragment of culture but an adjunct and, more dangerously, an obstacle to survival. The humanities, privacy, human rights, social safety nets, and an array of other cultural values then suffer the same fate.

Similarly, if time in the school day is limited, and if we are free to prioritize elements of a culture, then—even in rural Alaska—we put

English, American history, and American government first. Eskimo language, Eskimo dance, Eskimo stories, and Eskimo cultural ways and experiences become extracurricular add-ons at the end of the day, an adjunct to the "real" work that must be accomplished. When the budget crunch comes and both time and dollars for education are scant, then we drop the dance entirely to teach Eskimo or Indian children the "more important" fragments of nonnative culture. Those who do the prioritizing, who know what is a necessity and what is an add-on, are folks from outside the culture, people from Washington, D.C., or Indiana, California, or Texas.

The Future without Subsistence and Sustainability

Anthropology, the traditional humanities, and educational systems have all played roles in western European colonial expansion and have contributed to the fragmentation, breakdown, and oppression of traditional societies. Anthropology, however, has lately turned a serious corner that may send the enterprise in a new and more useful direction. Both Fienup-Riordan and Richard Nelson, another Alaskan anthropologist, began asking in the 1980s and 1990s not only what nonnatives can learn about traditional cultures but what we might learn from them.[5]

We are all one with the land—whether we choose to be or not. The greatest Western philosophers and scientists have not yet created a system that lets us stand apart, and they cannot, and they never will. The best Western science and philosophy and the traditional views of indigenous peoples all over the world are agreed on this. The notion that we moderns have somehow separated ourselves from Nature is fatuous at best and naive in the extreme. Modern science has made clear to us what traditional societies learned ten thousand years ago: that the chemical and spiritual bonds between us and the earth are absolute. Those ancient cultures knew what we are still learning, that when land health declines, human health declines, economic health declines, the quality of our lives declines—even to the vanishing point. We dare not believe that either technology or social engineering can let us off the hook. The question of subsistence and land use is not just a technological or structural one; it is an ethical and spiritual one. The dilemmas that subsistence presents are ours forever, and we have to work at their resolution forever. The only answers lie in us, in our cultures, in each one of us holding the subsistence worldview, living the subsistence life.

We are as dependent upon the land, and especially on its oil, as Plains Indians were dependent on the land, especially its buffalo. We have all read accounts of how wonderfully adept Indians were at using every part of the animal. Food, clothing, shelter, equipment, boats, household goods, needles and thread, bedding—all were derived from the buffalo. This is the testimony of historians, anthropologists, and other interested observers. General William Tecumseh Sherman was one such observer, and he knew that the more quickly the buffalo disappeared, the more quickly he could contain and control the Great Plains tribes.[6] No wonder the reports, as early as the 1850s, from Indian agents, scouts, and the Indians themselves that the buffalo were under grave threat went unheeded. We Americans knew we were exterminating the great herds and proceeded to our task with determination, skill, and thoughtful deliberation. We did not simply drift into slaughter; we knew what we were doing. This was not a thoughtless effort to kill buffalo but a calculated attempt to exterminate, or at least control, Indians. If we could make a dime in the process, so much the better; that was just good ol' American enterprise, to be lauded. And if it cost a species or two, more or less, it was worth the price.

So it is not hard to imagine an anthropologist or historian, five hundred years from now, describing early-twenty-first-century American culture. He will remark on the skill with which we used all parts of oil:

From some of it they manufactured clothing, from some shelter, elaborate transportation systems, warmth for their homes, tools, incredibly intricate industries, entire agricultural systems, even whole economic systems. But, alas, they were a one-resource culture, just like the Plains Indians whom they despised for their simplicity.

When it became clear how dependent they were upon that one commodity, they were at the mercy of the world, their lives and future controlled by those who could cut off their supply, or at the mercy of their own desires, incapable of restraint, refusing to husband the resource so it would last. As the number of their friends in the world declined and others turned their backs on them, a clear limit to the life of their culture became apparent. They waited, mesmerized and demoralized, for their end to come with the end of oil. They were just like the Indians at the mercy of General Sherman. They knew their history, but they thought it belonged to the past, not to them, and they did not learn from it.

Education for Subsistence

Imagine walking out into a glowing upper Mississippi evening in the spring. The blufflands above the great river are expanding into a green haze where the maples, anchored in steep hills, are just beginning to leaf out. The speed of that process thwarts the mind. In the morning the bud clusters of the red maple are sitting upright on the branches; by evening the red leaves are half unfurled, and by the next evening they are full, rich, and dark with red the color of dried blood; the bud clusters have been as "disappeared" as a dissident in Argentina. This land is lush, fecund, amiable in its appearance. Three months ago it was locked in snow, the wind a gale, the temperature minus thirty, the river white snow rippled like current over the hard ice. Everything that was to survive had gone to ground, seeking whatever cover the land or human shelter allowed. Survival is an issue here, for humans as well as everything else.

Or think of the arid lands of the American West, where the grass is shorter and the trees are reserved to stream banks and mountainsides, and the view opens and expands our vision. Everywhere one looks in the arid West, the land stretches out. Some of it still appears empty, devoid even of grass and game. Often the appearance is deceiving. There is game, but it is laid up for the day in the shade of a rocky outcrop, invisible to an untrained eye. Such desert land sustained American Indian people for thousands of years, though on occasion it failed them and brought hunger or even starvation. Either way it was a cherished landscape.

As one drives across a stretch of central New Mexico in July or flies over the tundra in Alaska in January, the land crowds the imagination to wonder how people manage to survive in such places. Yet they do, and they love that landscape as home in a more profound way than most mobile, modern Americans can imagine. Nearly forty years ago in Barrow, the northernmost town in Alaska, when the weather turned foul, it trapped some bureaucrats from Washington, D.C., who, after two days of whiteout and blizzard, were frantic to escape the godforsaken place. They were sitting in the Mexican café and bar, grousing and lamenting, when the door banged open and a young man in uniform, just back from Vietnam, blew in surrounded by a white dust of windblown snow. "By God, it's good to be home!" he shouted, while bureaucratic jaws dropped and local folks grinned and cheered. One might puzzle over how one learns how to live in and love such landscapes, but it is not learned in school. Such landscapes ask us, What sort of education is required to pass on the wisdom necessary for survival?

For most American Indian people, the primary purpose of education was survival on the land. That survival depended on learning an elaborate system of courtesy and reciprocity and gaining a profound understanding of complex relationships between humans and the rest of the world's creatures. One learned to shape one's behavior so that it would not offend the animals, plants, and streams human life depends on.

What have our own educational purposes become in our day? Is survival of the people still our aim, or are there other goals we strive for? Have we forgotten how dependent we are on the land for our people's survival, or have we slipped into behaviors that so alienate the land that in some not-so-distant future it will no longer support us? Worrying about committing offenses against animals, plants, and streams seems like primitive superstition to us, so we offend and diminish the world around us every day.

Our culture has learned much about how the world works, but we have failed to learn how important it is, how self-interested it is, to support the earth so that it remains vital and supportive of us. Now we lack the will to act on what we know. Crossing the land in our shiny cars or flying above it at 36,000 feet and staring down, we are faced with another question: What have we forgotten? We have forgotten that Ground Squirrel is grandfather to Bear. Austin Hammond's comment shows an understanding of the complexity of the relationships among the animals of a place and its human inhabitants, but our understanding of that place, and our own, is so poor that we do not know just what Austin meant. I think what Austin meant is that if we are not careful to treat Ground Squirrel with respect, it won't be just Ground Squirrel who balances the scale but something far larger: Bear. If we are open enough, perhaps we can let Austin teach us. Austin's story, taken as a metaphor for our relationships to the ecosystems we live in, offers a clear indication of our proper role in a place and the importance of caring for the small creatures we often overlook. If we harm the littlest critters, right down to the microorganisms in our soil, which we are killing off with our chemical fertilizers and pesticides, it will not be small creatures, Austin teaches us, but Nature itself that rises up to confront us.

In recent years people of the northwest coast and Alaska have worried much about the health of salmon runs. Incursions of foreign fishing fleets on the high seas, limited entry, and conflicts among sports fishers, commercial fishers, and indigenous people, all scrapping to save a portion of the salmon runs for themselves, are among the stories both local and national news media have carried.

The ancient American Indian idea that survival depends upon reciprocity, upon mutual consent between the hunter and the hunted, the earth's creatures and its human occupants, is an honorable concept—and we have broken it. In Naknek, Alaska, I heard from both elders and my students that in "old time," they were respectful of the little black creatures under the ground. They were always careful to return the whole skeleton of an early salmon to the spawning streams. This sign of respect was a ritual renewed every season upon the salmon's return. The salmon would then know that they were not ill used or wasted by their captors. When they returned to their homes, other salmon would ask, "How was it over there?" And they would say, "Oh, it was fine. They treated us well." Then others would say, "Good, we can go back there," and the runs would be assured.[1]

But in our time we have treated the salmon with contempt. We have used them as if they owe us their presence, and their death, simply because we want them. We have not honored them with rituals or been grateful for their abundance. Instead we have manipulated them, mismanaged them, ravaged and wasted them. We have not taken them for the sustenance of our bodies, a legitimate use, but for the increase of our corporate wealth, an increase that goes beyond any necessity for health to the stockpiling of profit beyond our capacity to use or spend.

The decline of salmon is more a reflection of us, and our character, than it is attributable to other causes. It is no wonder that there are fewer salmon. They should spurn us. It must be their grace and courage that enable the salmon to give themselves to us at all, an act of sacrifice made to help us recover our sense of awe and gratitude for the mutuality of all life, our awareness that we live and die by the grace of beings other than ourselves. When they return, we should be embarrassed in their presence.

Maybe they are only fish, and maybe the old rituals are only the signs of a superstitious past. Yet when we stand on the bank of the river teeming with salmon struggling upstream in one of the most ancient and epic quests to find a way home the globe knows, they appear mythic, more than fish, just as we are more than people only, both of us sojourners across a landscape of earth and river and time, participants in a larger, longer cosmic process, a journey that always ends in mystery. Both salmon and humans have a larger life, for we are both part of the seamless interdependence of all life, the give and take of existence, participants in the same process, all of us wanderers or exiles looking homeward. At a meeting of the National Congress of American Indians, Pat Locke, a Lakota woman, was exhorting a mutual friend, Bill Vaudrin, a Chippewa, to return to his reservation. He said, "Ah, Pat, I can't go back now. I've been gone too long; I'd just be a stranger." Pat, wisely understanding our larger human dilemma, said, "But Bill, we are all strangers, returning." What the people of old time knew, and what they taught future generations, was a system more gracious and more generous than we follow now.

Perhaps only when we reestablish the broken ties, when we atone for the dishonoring of our renewable resources, will the integrity of our systems be regained and humans once more be able to eat in peace, at home with our world and within ourselves. Then perhaps the salmon will return and offer us their forgiveness, and the ancient system of consent to our taking of life to sustain life may be restored. But how do we relearn ways to do that?

What ties exist between our educational practices and those of American Indians? The latter systems have been relatively successful in this place for thousands of years. Will ours work as well for as long? This is a critical issue for all of us, and especially for those interested in creating a sustainable culture.

Traditional Education

In the past nonnatives did not share much in terms of educational practice with American Indian peoples. Their methods apparently leaned toward observation of elders, cooperation, mentoring, and coaching, letting a child try something, more or less privately, until he was ready to reveal his skill. Beyond those essentials there was the use of stories, songs, dances. In nonnative American culture, especially since the movement toward consolidation, which took away older students' opportunity to mentor younger ones, methods have leaned toward individual effort, rote, competition, frequent testing, and display of knowledge, all in a system where songs, dances, and stories are often viewed as add-ons or frills.

What we know now of traditional Indian educational practices is tantalizing, but the possibility of our learning from them has lessened considerably after 150 years of forcing Indian children into our schools, assuming that our educational methods were right for everyone and that we Anglos had nothing to learn from local cultures. We expected them to learn from us, in a language they did not understand, yet we have not been educable ourselves. Those cultures have been in place perhaps fourteen thousand years or even longer. The real record of their longevity may well be buried under the rising oceans that have inundated our coastlines and erased evidence of their settlements in previous eras. Surely there must be something for us to learn from them about education and sustainability.

One of the disasters for Indian teaching methods has been the imposition of Anglo methods—in mission schools, Bureau of Indian Affairs schools, and more recently public schools—on the untested assumption that our pedagogy is superior. We dismissed their practices without examination, and a great number of them have been lost, our opportunity to learn from them largely forfeited. Even many Indian teachers and administrators, graduates of nonnative schools and colleges, who are now working under the constraints imposed by our alien public school systems, use our methods rather than their own traditional tribal methods.

When, on rare occasions, local Indian culture is included in the curriculum, too often the teaching methods and the context of the lesson remain those of the mainstream American classroom, not those of the local culture. It is hard to imagine that the inclusion of cultural elements apart from the culture's educational methodology will prove effective. Recently Indian educators like Rina Swentzell and Gregory Cajete and their many colleagues in tribally controlled colleges have been showing us how an Indian pedagogy might actually work. Their insights are important and, if we are educable ourselves, will eventually shape our own pedagogy, but for the present they are too little known and too little understood. Methodology, then, is not yet a common ground upon which we can build a system of value to all cultures and to the earth and its creatures.

Trying to find that common ground, my mind goes to thoughts of subsistence culture in Alaska, to subsistence as a base, as the culture from which education grows and for which education is essential. Subsistence culture carries within it the comprehensive complexity that faces us as we consider what a sustainable culture might be. Indeed subsistence may be an appropriate model for other cultures, once we understand what it really means. What do Eskimo people need to know to find their way in a subsistence world? And how do Eskimo people teach that?

Our Common Ground

We and our educators have forgotten that our nonnative culture is but a different elaboration of subsistence culture. We are all, regardless of differences in economies, dependent upon the land, as surely now as any culture since the world began. Is subsistence the arena, then, where educational systems as different as American Indian and Anglo-American can meet, our common ground?

We have something to learn from tradition at this point. Both our education and our environmental practices might be improved if we were to see ourselves as living a subsistence life dependent on natural resources, all of them interrelated, some of them in short supply, some of them threatened by our manufacture and use of chemicals and by the pollution and decay-resistant garbage we pile up, bury in our oceans, and send into the sky. Apart from the land, nonnative people will die just as surely as Eskimo people will, for we are equally dependent on the physical energy and nourishment the earth has to offer. But there is more to the parallel. We too depend upon the land for more than our food and energy. Our religion, art, philosophy, technology, and economy all come to us from the land. Whatever nourishes our spirits also depends on the land.

Whence do our own religions rise but from fire or earth, sacred animals and systems, stars or moons we hold, or once held, worthy of worship? In the Western tradition, Judaism, Christianity, and Islam were born and nourished in the great deserts. The desert tradition of hospitality, of patriarchs whose sternness matches the desert's harshness, still informs their codes of conduct, even now, after the religions themselves have spread to tropical rainforests. And where does Western philosophy find its root but from the examination of earth, from natural philosophy trying to determine first what is most knowable and then to extrapolate from that to the larger, more difficult unknowns?

In three thousand years, Western arguments about art have gone from how accurately it imitates Nature to whether it should imitate Nature. Nature remains a constant even when we avoid or distort its appearance. One question, often overlooked in discussions of the environment, has to do with how much we can alter the environment without altering our artistic expression. So much of our vocabulary and so many of the forms and images in our visual and literary arts come to us from the land.

Our economy, too, remains Nature bound. Logging, mineral extraction, and oil are contemporary examples of vulnerable "natural industries" whose demise causes widespread fear. Even our computers, rather than freeing us from Nature, put us at the whim of Nature's storms and temperature fluctuations. One-hundred-ten-degree heat never stopped my pen or typewriter, but my little Hewlett-Packard suffers, and its instructions warn me to protect it from such extremes.

Everything comes from the earth. One goal of education, then, one that will match what our lives require, whoever we are, is education for subsistence, a kind of learning that allows us to live in harmony with the world. This is a notion that is much more complex than it appears, for we must live in the world over vast stretches of time. We therefore must learn to understand and take care of the earth's health as we do our own, recognizing that they are inseparable. We may, with our fantastic machines, escape the earth, but what then? We have not escaped either the world or Nature. So how do we devise learning systems that assure us that we know how to live in the world with little enough damage to it that we have time? I am not thinking of brief time here but of transgenerational time: time for ourselves and other creatures; time for evolution to run its course in this place as it will in others for the unnamed generations of all the creatures we cannot yet see. What might such an education be about?

Restoring Wholeness and Integrity

Education for subsistence—rather than for jobs—would work toward the restoration of wholeness to our learning and lead us toward a restoration of wholeness in our culture, now so clearly fragmented among different disciplines with rigid boundaries in our schools. Our schools are still divided by race and its illusions of superiority and inferiority, by class and its myths of work and accomplishment. Division comes, too, from our myths of work—a "work ethic" that characterizes Anglos and a "laziness" that characterizes those of other cultures. We are divided further by our myths that portray artists as romantics and businesspeople as realists or pragmatists. So many of the distinctions we think we make rationally are built on the tricky sands of myth, self-deception, and self-aggrandizement.

Education for subsistence would assure the survival of the people, provide adequate definitions of who we are and what the nature of Nature is, and help us see the wholeness of that. One step toward that goal might be to rethink our view of nonnative cultures, to see them again as parts of Nature. Another might be to restore math, science, and the humanities to the unity they once had. This is a great, essential task for our time, and one that scientists are more apt, and perhaps better equipped, to undertake than humanists. At any cocktail party I more frequently run into a scientist who wants to talk about Wordsworth or Snyder or Kodaly than I encounter a humanist who wants to talk particle physics or chaos theory. Education for a subsistence life would alter many of our definitions of who we are and how we should live. It would also radically blur the distinctions we make between disciplines.

How much of education, of learning, is a matter of acquiring definitions—of mulling things over in our minds till new definitions emerge, rather than accepting the classifications schooling puts on us. We have acted as if we are not subsistence people, or we have forgotten that we are, but the concept of subsistence allows us to recognize our culture in the same way native peoples recognize theirs, whether we acknowledge that fact or not. In 1991, J. Edward "Ted" Chamberlin, professor of comparative literature at the University of Toronto and former chancellor of New College there, sent me a note about the importance of such recognitions in education:

It is not much as things are now, but—(1) any act of classification, both of disciplines and within disciplines, is an act of definition . . . an act of naming, with all of what Merwin would call its imperial implications. And the system we now support is obsessed with classification—as distinct from and often opposed to the sense of wonder. (2) We like to think of literature as helping us to live our lives. To the extent that this is true, it must in some sense help us develop definitions different from what I refer to in (1): i.e., more positive, though still establishing limits of self, a community, etc. At the same time, I don't think we teach literature that way, at least normally, in the current system.

Should classifications and definitions—knowledge—be what learning is about? Or, in light of Ted's comments, do we want to foster recognitions and tentative ideas that we can test instead—those awarenesses that do establish perimeters but lack the distancing and denaturalizing we have come to fear and abhor when carried too far? Recognitions lead us to think beyond wherever our mind has previously come to rest. They lead us to wonder, to speculate, to brood, and to create for ourselves rather than accept without question the opinions of others. Our minds tend to stop when we have definitions; wonder tends to cease. We think we know a thing when we have defined it. But recognition, re-cognition, by its very lineage as a word, allows for continuing both thought and wonder—two important elements for keeping the educational process alive.

Why do the data about the destructive impacts we've made on the environment appear to fall on deaf, or at least indifferent, ears? Even folks who recognize the threat may remain unmoved because they have not made the prior recognitions necessary to change behavior. Perhaps the majority still have not recognized that we are indeed a subsistence culture, dependent entirely on the land. Or they have not recognized that we still languish in the legacy of Newton and Bacon, thinking that we can separate our science from our humanities, and in the legacy of Descartes and Kant, thinking that we can separate ourselves from Nature. Those two separations, it seems to me, go hand in hand; it is a short step from the former to the latter. (Ah, the ease of the second guesser, the Monday morning quarterback!) Those separations are also the source of Philip Sherrard's notions of the dehumanization of man and the desanctification of Nature. The shallowness of scientific thought, in Sherrard's view, is ultimately caused by its (our) separation from traditional wisdom.[2]

But if our knowledge of the world can be split off from our knowledge of ourselves, why can't we separate ourselves from the world? Our minds take such turns the way salmon occasionally take a turn that leads up a fatally dry tributary. If we can be separate from Nature, superior to it, able to control it, and if we are not a subsistence culture or do not recognize ourselves as such, then the consequences of our destructive acts do not matter much. We can assume we'll find a way to fix things up again and "manage" Nature back onto the "right" path. But if we see ourselves as a subsistence culture, dependent on the land for all health, then—perhaps—we will recognize that we cannot destroy the land without destroying ourselves, and we will change our behavior.

One essential task for an educational system that will buy us transgenerational evolutionary time and allow us to continue to subsist is to teach us how to find our way toward reconnection. It must recognize the conjunctions between ourselves and the world, between what we know of the world and of ourselves. Subsistence-based education will enable us to take steps toward recognition of coherent patterns, to take steps toward wholeness and toward the kind of cooperation, rather than competition, biologists now say runs the world. The wholeness and the cooperation are there, present in the world, fundamental to it, operating in it. We do not have to create it. We only have to let our comprehension grasp it, our recognition apprehend it, for us. If we and our children can come to know the world in this way, we can live in the world over time, and the earth will be viable for every generation.

The task is formidable, for our science still remains largely a separate enterprise. It remains a reductive process of taking things apart, removing them from their contexts, assuming that somehow we will find what things really are by looking within, finding what is smaller and smaller and therefore, we think, closer to the "core," as if the core must somehow also be the essence. That is perhaps half of the true task of science, but the last few hundred years it has been the major part, perhaps because it has worked so well for us. But what if everything is onions? What if there is nothing at the center of our reductions? What if the real meaning of our search is to be found in relationships, in the scent of the connections between layers? What if the essence of everything lies not inside itself but outside, in its connections with other things? The new science of ecology is looking hard to discover relationships and how they work, but at present it represents only a small part of our scientific endeavor.

To find out if connections are the real essence, we need to look at wholes and relationships, not only within but at and between, trying to recognize what exchanges are flowing. Then we might discover how they and we reach out, are stimulated, moved, made operative, by things outside ourselves. It seems we have only begun to develop the skills for that kind of inquiry, that we do not yet know how to look at wholes or at larger systems rather than smaller. Perhaps our astronomers have that capacity, but it hardly registers on the popular consciousness. Astronomy now is more apt to be equated with NASA than with science, though even there, what captures the lay imagination is the technology, the power of the rocketry, the courage of astronauts, the miracle of huge telescopes slowly tracking stars without a human hand to guide them.

Even in our psychiatry we are asked to look deeper and deeper within. Only recently have family therapy and twelve-step groups made an effort to look between, not exclusively within, examining our relationships to discover how we really feel about our lives and our relationships with others. What many have found, looking at relationships, is life and the means to endure it, even to love it. Perhaps one distinction to be made between psychiatry and a shaman's practice is that the shaman looks first to an individual's relationships to the land and its creatures and to the tribe, seeking to discover what is out of balance, lacking harmony. The psychiatrist, in contrast, looks first at how an individual sees his or her own history, or simply prescribes the latest medication. Emerson says, "I cannot greatly honor minuteness in details, so long as there is no hint to explain the relations between things and thoughts; no ray upon the *metaphysics* of conchology, of botany, of the arts, to show the relation of the forms of flowers, shells, animals, architecture, to the mind, and build science upon ideas."[3]

Maybe the ancient astronomers were right: we can come to know ourselves best by looking to larger systems. Coming to recognize ourselves as part of ever larger systems or patterns, we may finally find the pattern for ourselves in the linkages between systems and, finally, in the stars—not by consulting astrological charts but by achieving some Chaldean recognition that in the turning of stars we ourselves turn. We are learning that even our own tiny turnings and flounderings have their influences too, clear out to some already extinguished nebula whose light is still pulsing toward us. We have known from the beginning, if the elders of every culture are right, that the exchanges we need to understand are not only between us and kin, neighbor, countryman, or stranger but also between us and stone and plant and field mouse; between us and ground squirrel and bear, and everything in and beyond the galaxies we know. If that is so, we have to be careful how we move on this earth, how we move through our lives. If we act as if we are disconnected, the connections will surely snare us, for, as Gary Snyder reminds us, they are nets, not tethers.[4]

If we break the connection between ourselves and the world, then we are merely fools who may unwittingly jostle the stars from their courses. Kathleen Dean Moore, in *The Pine Island Paradox*, asks a question that reveals how dangerous that might be for us: "And what is a human life but a rearrangement of molecules that once were stars?"[5] Enabling us to know the world, to know ourselves and our place in the midst of it, and to know it whole, could be the purpose of an educational enterprise worthy of people who, like Alaskan villagers, other indigenous peoples, and us, are participants in subsistence cultures. Fulfilling that purpose through our educational systems, finding our common ground in the world, may ensure the survival of all cultures.

Sustainability

———— ● ● ● ————

Education for Sustainability

THE BEST EDUCATION DOES NOT necessarily come from the best teachers or the best schools. Behind those happy elements lies an important factor not often addressed in discussions of the problems of public schools: a desire for self-cultivation, the self-discipline that insists on learning whatever the learning environment might be. Discipline is a constant in our conversations about schools. Self-discipline doesn't often register on the radar in such discussions. In my experience, teachers most often talk to students about obedience to them and to the rules. I don't ever remember a teacher talking to us about the importance of disciplining ourselves. How do we develop in our children a desire to take their own discipline in hand? I don't have an easy answer for this. My own self-discipline waxes and wanes. The trouble with self-discipline, at least mine, is that it has no memory. That I had enough discipline to pull off one major, demanding task is no sign that I will marshal the discipline for the next. But a sustainable culture without self-discipline is an oxymoron. What is the role of education in creating a sustainable culture? What is the role of education in creating a self-cultivating, disciplined citizen who sees his or her role in the culture as serving family, community, nation, the great world of ten thousand things? How do we create citizens who are willing to place limitations on themselves, their power, and their acquisitions so that there might be enough for everyone and for those other creatures whose lives humans depend on though we may not recognize it? How does education fit in the issues we have been addressing here? There are other questions that swirl about education in our time that also affect our ability to create a sustainable culture. They have largely to do with the purpose for which we educate.

A Small Survey

The purpose of education, according to some Native Americans, is to ensure the survival of the people. Koyukon Indians along the Koyukuk River in Alaska—and many other indigenous peoples around the globe—use traditional stories to teach their children those values that will maintain balance and harmony in the world. The Koyukon call these tales "true stories" and say, "They teach us how to be human."[1]

In the Western tradition, Socrates wanted to educate systematically so that the republic would comprise citizens trained in virtue. He seemed to believe that the young already know all the knowledge the world has ever contained and that their education can be achieved by reminding them of what they have forgotten. Others see children as empty vessels into which adults must pour the information they have learned so they can lead satisfactory lives. Socrates requires us to choose one view or the other. Our purpose becomes either to draw out or to stuff full, a choice that requires us to adopt radically different teaching strategies, a fundamental choice now largely ignored in schools of education. Means and ends, as Socrates recognized, are clearly linked.[2]

In the Chinese tradition, Confucius believed that good government becomes possible only when men gain that clarity of thought that comes from the discipline of finding and using clear language. They clarified their thoughts by searching the limits of knowledge, educating themselves as widely and deeply as possible, and insisting on the accurate word. The first task of government, Confucius claims in his *Analects*, is to "rectify the names," to call things by their right names. Self-cultivation, or self-discipline, was seen as a lifelong task and the key to achieving wisdom. Having disciplined themselves and brought order into their own lives by clarifying their thoughts and their language, men could bring order to their families. Having brought order at home, they could bring order to the society.[3]

In ancient Rome, Seneca, writing to his son, decries scholarship that strives to tell him exactly which rock Ulysses landed upon but cannot tell him how to quiet the wandering of his own heart, or the mathematicians who can measure exactly the extent of his landholdings but cannot tell him how much is enough. In essence Seneca asks us to think not only about the purpose for which we educate but about the relationship between scholarship and the humanities. Apparently, in Seneca's view, they are not contiguous, and the amount of land one holds is not only a practical matter of survey mathematics, accounting, and the law, but a matter of philosophy and spiritual health.[4]

In our time and culture, the purpose of education is to get a job and make America the number one economy in the world. To assure this we create apparently never-ending sets of standards to measure our children's achievement and then hand them standardized tests to make sure they are meeting our goals for them. Contemporary schools must graduate students who can accomplish what an older, purportedly better-educated generation could not.

Keeping my survey small, I asked my son Kevin, who works in an alternative middle school, what the purpose of education is. He stopped working out for a few minutes, thinking about the question, then said, "I think it is to create openness." "Openness to what?" I asked. He thought again and said, "Balance and harmony," smiling, because he must have known what my next questions would be. "What are you balancing?" I asked on cue. "Mind and heart, the rational and the emotional," he returned. "And the harmony?" "I think it's harmony within ourselves, and between ourselves and others, and between ourselves and the natural world." He began working out again, but I couldn't help interrupting to ask, "Openness to anything else?" He stopped awhile, staring off, then replied, "Openness to becoming natural in ourselves instead of contrived." He meant, he explained later, that we have to cultivate ourselves in such a way that our selves grow naturally into lives that we do not have to edit or be embarrassed about or protective of, that do not require secrecy to maintain a facade of probity. Respect, caring, compassion, thoughtfulness, and openness would become our natural life.

Kevin is as eager as anyone I've ever known to learn things: how the world actually works, what ideas science and history and philosophy can give us to consider, what the world's greatest thinkers have discovered through the ages. Yet for all the information he has acquired and the knowledge he loves to mull over, his learning seeks still larger aims.

John Goodlad, director of the Center for Educational Renewal at the University of Washington in Seattle, and his colleagues Roger Soder and Kenneth Sirotnik hold that education is moral in nature and that a moral life is therefore at least one essential purpose of education.[5] In this they echo Socrates, who insists that we educate to instill virtue. Neither Goodlad and his colleagues nor Socrates has any interest in an education that is merely moralistic. Virtue and morality lie in citizenship exercised for the common good—not in a canon of laws that govern personal behavior but through participation in a community's life that can grow only from persons whose purpose in life is to achieve wisdom. Acquiring wisdom is the ultimate purpose of education for Confucius, Socrates, Seneca, Goodlad, Soder, Sirotnik, and Kevin too.

Some Questions

Do we and our educators have the same purpose in mind for the education of American Indian, black, Hispanic, and Asian students as we do for white children of the upper and middle classes? Is our purpose in educating women the same in every respect as our purpose in educating men?

Despite the hue and cry about dead white males, Plato was the first great feminist we know of in the written tradition of the West. In *The Republic* he insists that "if we are going to use men and women for the same purpose, we must teach them the same things." Plato also insists that women and men have exactly the same capacities, except, perhaps, physical strength. Even then, he points out, some women will be stronger than some men at even the heaviest physical tasks. Both the means and the ends of education ought therefore to be exactly the same for men and women, Plato thought, including training to become the philosopher king.[6] Do we agree in our time? Is our purpose as coherent as his?

Can we educate to good purpose and quantify or objectify every-thing in the process? Why do we want our education to be so objective, so quantified? Life does not give us objective tests, so why do we spend so much effort teaching people to take them? Life is the least standard-ized, least objective, murkiest, and most muddled test we'll ever take, with no clear answers to any of our most human dilemmas.

Is it appropriate to strip the teaching of values from our schools? Is that even possible? We have tried to prohibit discussion of values in our public schools for some sixty years. Why are we surprised, then, that our prominent citizens are going to jail and their children and grand-children, who are now our students, appear to have no values beyond self-interest or are, fortunately, learning their values from sources other than their parents? We scorn the McGuffey readers for their obvious touting of values and prefer our own, more subtle and objective texts that we claim, deceiving ourselves, to be value free. Why do we think that prohibiting a discussion of values in schools is not in itself a pro-found declaration of value, questionable as it might be?

Do we educate for each person to develop a self to the greatest pos-sible extent, educating toward maximum individual growth and self-expression? Or do we educate for participation in a community where self-restraint is a primary virtue and citizenship a cherished responsi-bility? Do we try to get both these results from a single system? Sam Keen holds that "education has two primary foci: it must initiate the young into the accumulated wisdom and techniques of culture, and it must prepare the young to create beyond the past, to introduce novelty, to utilize freedom."[7]

What is the nature of the person? Is the self unique, private, to be developed alone? Tribal cultures that have already had longer lives than Western culture seem to be among those in which the sense of self is not primary. It may be that the stronger the tribe feeling is, the weaker or less important the individual self-image is. In Alaska many educa-tors believed that native students suffered from a poor self-image. But if the sense of tribe comes first, and if it is strong enough to outweigh a sense of individual self, is there a self-image to damage? It may be that in tribal cultures the self-image is not suppressed, submerged, or damaged; it never existed in the first place. Is it possible that the image that has suffered is instead the Indian child's tribal image, and it is that image that needs strengthening? That seems possible given the effort that our educational and political systems have invested in destroying the organizing principles of the tribe. The problem Anglo educators often ascribe to American Indian kids on reservations, then, may be a reading into their plight from our own perception of ourselves, and not their real condition at all. How can white schoolteachers, brought up in a culture that does not understand tribalism and generally has been hostile to it, possibly be competent to undertake restoring a tribal image?

Could such a circumstance also occur among the street gangs our urban citizens so greatly fear? If we insist that we need to improve their self-image, are we saying that we need to improve what does not exist for a gang member? If membership in a gang provides the primary identity, how can a single member have a poor individual self-image? Is the desire to be part of a gang a search for community, a desire for freedom from the exile, loneliness, and fear that grow out of America's rampant individualism, its isolation of those who are different, poor, or otherwise marginal? Could it be that gang members are more civic minded and intelligent than the rest of us, unwilling to remain outside a community to which they can contribute and in which they can find support? And given the questionable state of middle-class, white, Western culture with so many problems growing out of its fixation on the value of the individual, how can anyone be sure that replacing a tribal or even a gang identity with an individual sense of self is such a good thing?

Both Westerners and the Chinese desire the development of the self, said Ron Scollon, a linguist who has lived and taught among the Chinese, "but in the Chinese view, the self can never be developed alone but only in conjunction with others." In a telephone conversation, he pointed out that Westerners seem to believe that rationality, the ability to reason, is the shared common characteristic of humankind, the one capacity that sets people apart from other animals, and that sets one individual apart from another. "But the Chinese," Ron declared, "would say that kinship is what we share." Indeed, "Deliberate moral reasoning, in that view, is a sign of inadequate personal progress rather than a distinctive feature of humankind."

The Purpose of Contemporary Education

When asked about our American culture's apparently muddy notions of the purpose of education, one English professor recently said of higher education in California, "Every university or college president I know in this state is very clear about his purpose as an educator. . . . Each one sees his task as becoming number one in a competitive system, or at least among the top ten in the region. And they know how to achieve that too—by quantifying everything. The college with the most Nobel laureates on the faculty or the most Pulitzer Prize winners in the English department is clearly the top school." Some public school administrators also remain clear about their purpose. That superintendent of schools in Alaska who insisted that his task, through the schools in his rural district, was "to wipe out the last vestiges of Eskimo culture as fast as I can" clearly knew his purpose. But it was his purpose, not that of any Eskimos in the villages he was hired to serve. Thinking through our purpose in schooling is critical, for if we are not clear about it, education may not lead the society in directions it expects. There is no mandate in our schools to see that one outcome of schooling is a sustainable culture.

Speaking in 1992 as part of a panel at the Island Institute on the purpose of education, Ted Chamberlin made explicit some of the ramifications of educational goals and pointed up the complexity and the contradictions that lie within our notions of purpose. He noted that throughout the history of education, whether in Europe, Asia, Africa, or Australia, one purpose of education has been "to prepare individuals to meet the challenges of the world, and to satisfy its demands." He described this as a conservative purpose, "one of sustaining the values and the meanings of society." Another purpose in our educational systems, he continued, has been "to separate individuals from the world for a period of time, and provide them with an opportunity to strengthen themselves to challenge the orthodoxies of the world." This latter purpose is seen as preparing the young to look at the culture critically, "with the purpose of changing those practices, those givens, and changing as well the theories which sustain those practices." This latter notion also "presumes some idea of progress, some notion that things can be improved."

The educational system contains two fundamental contradictions, according to Ted. The first has to do with deciding whether our purpose is to focus on improving the individual self, "independent of the priorities or the prerogatives of the society," or on improving society "by creating more civilized individuals." The second stems from Plato's questions in the *Meno*: Do we seek to draw out of the individual what he or she already contains, assuming that there is a "best self" lurking in each of us and that "the educator's task is to lead that self into the light"? Or do we "assume that each individual needs something from the outside, something that is not inherent; that we need to lead the person out of ignorance, stupidity, or barbarism, and that we can do that only by providing the person with information, values, or ideas that are now missing"? Ted pointed out that "the second notion is what informed much education during the nineteenth century, especially in the European empires: in Canada, in India, in Africa, in the West Indies." He described the imperialism inherent in such a notion by showing how it took root in "a system that held a condescending view of indigenous people. And it had its influence on the developing of education in the United States, especially in the development of education from what were often imperial metropolitan centers, which, of course, knew what civilization was all about, compared to the colonial hinterland." He concluded, "So education was often determined in the metropolitan centers of the East, and then sent out West in order to civilize the other areas of the country."

This latter circumstance is precisely what American Indian and Alaska native students have had to contend with since schools were imported and imposed on them to assist in their "civilization." The outcome of such a purpose was supposed to be assimilation. The real outcomes included the denigration of local cultures and a number of shocking losses: loss of vernacular language, loss of traditional wisdom, loss of a unique worldview, loss of useful subsistence skills, loss of confidence, loss of the organizing power of the tribe, perhaps even loss of the tribal psychology that it created in its members, and loss of the ability to find one's way in the world. All those losses reinforce Ivan Illich's argument that the humanities have been waging a long, cold war on subsistence and the vernacular—upon indigenous peoples.

When we think about the purpose of education, we need to stress the singular. Robert D. Arnold, former executive director of the Center for Equal Opportunity in Schooling, speaking as another participant on the same panel, emphasized the need for "purpose, rather than purposes." The plural may be more inclusive, he claimed, but it exposes us to long lists. "The singular will let us more easily distinguish between purpose and goals and objectives." The latter two terms have been used widely since the late 1950s as part of the development of governmental planning processes modeled on industrial planning efforts. They were applied to education with considerable determination in the mid-1960s and 1970s. Both terms are severely limited. They express something far less than we need to discover. They turn schooling into an assembly line over which others have control. "Here, take this part. Now take this one. We'll assemble you according to our pattern." What a dismal future that portends; only a stifled creativity and lack of innovation are available for tools to sustain our culture. As used in Bob's deliberations, "'Purpose' is a broad or global statement of intended outcomes" and "could represent a community's highest aspirations for its children and youth." He continued, "A clear sense of purpose will help identify narrower outcomes called 'goals' or 'objectives.'"

Goals and objectives are not a purpose, and their presence will not make up for the absence of purpose, in Bob's view. Objectives may be important for the military, business, and industry; their value for education is doubtful. Objectives are closely hewed to obliterate any hope of serendipity, the unexpected, the unplanned, the unpredictable, and ultimately the most pleasurable learning. Some teachers learn quickly that what they thought they taught was not what their students learned. Students often find that what they learned that was most significant to them is never covered on a test; it may even have been reached by an intuitive leap away from the lesson being taught, an individual but not classroom epiphany experienced while brooding or staring out the window. Further, teaching to achieve educational objectives fragments learning and makes any holistic view of the world even more difficult to achieve than it is inherently.

National consideration of our purpose in educating our people has been too long in coming, but a national declaration of purpose made by government is always a travesty and inevitably becomes bungled when translated into policy and implemented locally. Perhaps a consideration of purpose becomes possible if educators, other intellectuals, and the public try to think it through in conversations, but it will happen best at the level of local school districts. Out of those myriad conversations there will surely be some common ground, statements of purpose whose values, expressed or implied, clearly overlap. Anything less than such a community effort is doomed to fail. A whole series of "reforms" in schools that were put into practice from 1965 until 1980 have left many people still deeply unsettled about our educational programs. Reform is clearly not enough.

A statement of purpose should grow out of conversations among our citizens, not be handed out by an individual or left to professional educators or politicians. My notion, assuming that we have a conversation going here and that I can participate in it, is that we educate for life, and not for mere survival but for a particular kind of life. We educate toward an authentic existence, rooted in the deepest nature of Nature, creating a fullness in ourselves that allows us to give of ourselves to others. We educate for a life that is concerned for the community, the state, and the earth and all its species and geologic forms. We educate for a life that reaches out in appreciation, gratitude, and love to the very stars. Yet the purpose of education in our time often seems to be defined by the needs of the state or the corporation or the economy rather than the needs of people or the community, and certainly not for lesser species or the earth. Instead of asking ourselves what our people need to know to prepare them for a future that defies prediction, we ask ourselves how to train people to ensure the state's economic development or to meet the need for workers in a specific industry. Perhaps we will never educate anyone very well until we disentangle education from the economy, until we divorce schooling from getting a job.

We all come from traditions that once were clearer about the end of education than we are today, when training for a job (probably in transition as we train for it) seems to be our primary goal, and achieving satisfactory scores on a standardized test and behaving correctly in social situations are common and acceptable intermediate goals. Some pursue such goals even though they are not the purpose of education and may not be even distantly related to it. Meanwhile the purpose for which we educate remains ill considered by some, unconsidered by many, and poorly defined by most.

Wisdom and Survival

It begins to seem as if the people and cultures of old, trying to educate for survival and for intimate knowledge of the world and our relationship to it, were wiser than we. Their view is a long one: a life aimed at intellectual and emotional growth and increasing self-discipline and service to the community. Our view tends toward the immediate return: an easy job with benefits. Their cultures have survived for thousands of years longer than ours. Who has a greater sense of purpose? For American Indians, survival of the people, including creatures other than humans, was not only the ideal but was seen as realistic and worthy of pursuit. Western culture's traditional interest in educating for citizenship, for virtue, and for the common good seems to offer a parallel purpose, though one often lacking in concern for the common good of nonhuman creatures. By comparison, our contemporary goal of individual accomplishment, defined as skills for the job market and ever increasing compensation, and our national goal of making America the world's number one economy, seem hollow and thoughtless. Despite our books and our science, our technology and our access to information, we no longer appear to have much capacity for thinking hard about difficult issues or to have very high expectations for our schools or, tragically, our young people. Our capacity for respect, for being fully present to others and to the world, is not widely expressed now. It seems we prefer to display our power rather than our respect, to get others to pay attention to us rather than be present for others.

Though there has been much national ferment in education since the 1960s, it has been a long time since we as a nation have carefully examined our presuppositions about education. The principles espoused by Mann, James, and Dewey, and in the great debates about the nature of education in the 1930s, have borne their fruit. We may differ as to whether it is bitter or sweet. Critiquing those earlier debates easily leads to bashing the results, and although that is a particularly sour fruit, many seem to relish it. It is more important, in my view, to think hard again about our purpose, not limiting ourselves to a critique of our recent past or our apparently numbing present but exploring all the sources of thought that could be useful from every era and culture. The inquiry itself is useful, and the implications of the outcome are important to every citizen in a democracy.

If we have a clear vision of our purpose, then we have a better chance to accomplish something worthwhile in our schools. "If we have no purpose, or cannot articulate one, then there will still be a purpose," Bob insisted, "but it will be by default," and it will be revealed only by the products of our educational systems. Then if the product is not to our liking, it will be too late to alter either our purpose or our means to achieve it.

Whatever we decide about the purpose of education in America, Bob and Ted make clear that we must come to grips with some clear purpose and we must trace the implications of that purpose as far as we can, so that we will not get results we did not anticipate. Even for our sophisticated culture, the clues we have from traditional peoples may drive us toward wisdom and survival of the people—and other creatures—as appropriate goals for our difficult and dangerous time. Only when we know the purpose can we devise the means to achieve it. "The end does not justify the means," Sufi educators say; "it provides it."

Imagining Sustainability

PERHAPS THE DESIRE TO create a sustainable culture begins with wonder that we are alive, participants in this most complex system of flesh and stone and stars and ideas and spirit that surrounds and sustains us and within which we "live and move and have our being." Or perhaps the imagination is simply prompted by the same question Mary Oliver raises in her poem "Spring": "There is only one question: How to love this world."[1] How *do* we love the world—with all its frequent beauty and inevitable pain, its occasional gentleness and nearly ubiquitous violence, with riches for a lucky few and poverty for most, our peculiar, personal mix of creative impulses and destructive tendencies, and all those losses that mark our every day? How do we learn to care enough for the wonder of it that we finally take care of it, begin to create and nourish a sustainable culture, one that recognizes that only by sustaining the rest of the world can we sustain ourselves? How do we learn to see it whole, to recognize the unity that lies beneath all its different aspects, that unity our eldest elders say knits all things together, despite the fragmentation, borders, boundaries, and ghettos that surround us?

It can be a tough world to love: Government and politics gone nuts with violence as the first response instead of last resort, and a public rhetoric of deceit-fueled hype rather than serious discussion. Education aimed at pitiful work skills, outdated in three years, that doom our children to meaningless jobs so the wealthy can get wealthier when we instead need, desperately, to generate the wisdom that may save us all. We work our farms and factories till our backs are bent, or we sit at our corporate machines that do not open up new worlds for us, and we stay there before a pale blue screen till our wrists ache and fingers won't straighten, and we never really seem to get ahead. Tribalism and poverty, disease and drugs crouch at the bottom of the iron gates behind which lie communities of splendid homes, and there is no real safety on either side of the gates, or anywhere.

A couple of years ago, I was facilitating a meeting of medical practitioners who were plotting against chronic diseases in our little town. They were bright and caring, and it occurred to me that they might really be able to eliminate, or nearly eliminate, debilitating diseases that destroy the pleasure some older citizens might find in their closing years. What came to my mind was the question, What's wrong with this picture? Suppose we did have a town that was perfectly healthy physically. Would we have a sustainable community? Probably not. If we had physically healthy communities clear across the state, would we have a sustainable state? Not likely. If we had perfectly healthy states clear across the nation, would we have a sustainable nation? No again, and the reasons are the same at every level.

In our town we have a long history of racism. White people have not cared much for Indians since they settled on Indians' land, took it for their own, and built the town. Perhaps we don't like them because their presence on a small reservation at the edge of town reminds us of that. In recent times our largely German and Scandinavian white population has seen some color seep in—hardly a trace at first, but now a discernible pattern: blacks, Vietnamese, a few Cuban families. In our community there are some folks who do not see this as an opportunity to grow and learn and extend our pleasure and insight into how the world is composed, to enrich the tapestry of the present and create a richer tradition for our children. Instead we see it as a shadow, a vague threat. This view extends clear across the state, and throughout the nation.

Early one morning I was standing outside the train depot, waiting to board an Amtrak train to Chicago. Only one other man was there, an old-timer who started a conversation. "Come down often?" he asked. "No," I replied, "just heading for Chicago to meet my wife." He often came to the depot in the morning, he said, just to watch the trains and see if they were on time. He was retired now, and it was nice to get out in the mornings—"gives me some exercise and something to do." He'd worked as a welder for forty-five years, and now he was done. Glad of it; "got tired of getting burned." He'd come here thirty-five years ago and stayed. "It's a beautiful little place," he said, "and folks are nice." He seemed like a pleasant fellow, a good neighbor type himself. He spoke some more of the virtues of the place, then added, "I even like the cold winters, even that thirty below cold we get sometimes, you know. Know why? Them there African Americans, they call themselves, they can't stand the cold. Got that cold last winter and one of them, a couple doors down, couldn't take it, just gave up, said he was going back south. So I even like the cold."

Some of "them African Americans"—and now some folks from other countries—live in low-income housing. Some white people assume they are involved with drugs. Others assume they are taking our jobs, even though we are employed and not looking. If they drive a newer pickup than ours, some feel that somehow they've cheated us out of it; it's our country, after all. Some assume violence. But there is far more white violence against minorities than the other way around. And some also assume, without a hint of a fact to support the assumption, that *those* parents don't really take good care of their kids, that they do not support the school system with their tax dollars, that they are somehow a drain on us and on our resources. And many of us are uncomfortable if we overhear a conversation in passing, one that does not even involve us, that is in a language we do not understand.

And those Indians on our outskirts? They are doing all right for themselves now, have a prosperous casino, seem to be using their new funds pretty wisely, even put some of it in our local bank and support local causes we care about too. But if we don't like poor Indians, we seem to like prosperous Indians even less. There is resentment when their children ride to school in a new car. As one wise elder in town commented, "Well, I never heard anybody complain because the doctor's daughter drives to school in a new BMW," but he seems distinctly in the minority, his more generous view not widely shared.

Our town isn't much different from most in our state. Somalis have been attacked in cities nearby and beaten severely, even killed. And our nation's record in regard to race is dismal. Immigrants rarely find a welcoming home here; immigrants of color fare even less well. Till recent years the signs in our merchants' shop windows ranged from "Irish need not apply" to one I saw in Alaska that read "No Indians and no whores." The signs finally came down, but the attitude behind them has been harder to change. Until it changes in our town, our state, our country, we will not have a sustainable society. Sooner or later it will fracture in violence, and only smoking shards of our culture will remain, or we will become deliberately, compulsively repressive and oppressive so we can maintain "order." If we do not diminish our own national disdain for others, others will do it for us, and our military may not be able to withstand the onslaught.

Our fear of those different from ourselves takes other forms as well. In our town finding a place for those whose sexual orientation is not absolutely heterosexual has been difficult. Gays and lesbians have often faced slurs or job discrimination—"the slings and arrows of outrageous fortune"—that make one marvel at their grace and openness in the face of public sentiment against them.

We could have a physically healthy town, and we *might* be able to fix our racism and our homophobia, yet there are families of grand wealth in our little community and others who are not making it, and more than a few who are marginal, making it but just barely, who could fall into a poverty nearly absolute. That's not much different from other towns around the state either. And the nation has grown increasingly hostile and unwilling to work on the issue of wealth inequality in recent years. Who's really ready to solve the problem? Are local merchants and businesspeople ready to reduce poverty in their communities? No, not as long as they are unwilling to pay wages that are above poverty level. Is agribusiness ready to reduce poverty? No, not as long as it continues to rely on imported workers and pay its employees wages far below poverty level—and finds ways to reduce that amount—while denying them a safe and clean place to live. And not as long as it continues to play to commodity markets rather than practice sustainable farming. Is Congress ready to reduce poverty? No, not so long as it sets a minimum wage that leaves working people below poverty level and is designed not to put money into the pockets of the poor but to protect businesses. How can a congressman keep a straight face talking to his constituents about reducing poverty when Congress has to debate for years whether to raise the minimum wage? How serious are we about creating a sustainable culture if the wealthy are unwilling to reduce their own income 5 percent in order that those who work for them can receive a 20 percent increase in their income?

Henry George, the now little-known antipoverty warrior, speaker, and author, had it right back in 1885, even as he lifted part of his speech from Leo Tolstoy:

Consider the matter, I say it with all reverence, and I merely say it because I wish to impress a truth upon your minds—it is utterly impossible, so long as His laws are what they are, that God himself could relieve poverty—utterly impossible. Think of it and you will see. Men pray to the Almighty to relieve poverty. But poverty comes not from God's laws—it is blasphemy of the worst kind to say that; it comes from man's injustice to his fellows. Supposing the Almighty were to hear the prayer, how could He carry out the request so long as His laws are what they are?

Consider—the Almighty gives us nothing of the things that constitute wealth; He merely gives us the raw material, which must be utilised by man to produce wealth. Does He not give us enough of that now? How could He relieve poverty even if He were to give us more? Supposing in answer to these prayers He were to increase the power of the sun; or the virtue of the soil? Supposing He were to make plants more prolific, or animals to produce after their kind more abundantly? Who would get the benefit of it? Take a country where land is completely monopolised, as it is in most of the civilised countries—who would get the benefit of it? Simply the landowners. And even if God in answer to prayer were to send down out of the heavens those things that men require, who would get the benefit?[2]

How sustainable is our little town, or our country, if the wealth of a few increases dramatically while poverty overtakes many? The center cannot hold under those strains, and we can't last long. Is it true, as it seems, that our real values as a nation, as members of our community, and as individuals lie in simply increasing our personal income, even at the expense of others?

But even supposing that we could find more equitable ways of distributing wealth—admittedly a long shot—and that our physical health was not problematic, and that we were able to temper our racism, we still would be drawing down natural resources and energy faster than the earth could renew and replenish itself. Near our town we have some agribusiness that depends on toxic chemicals. The size of corporate farms is still growing, and corporate farms demand chemicals. As one such farmer said, "I suppose that chemicals aren't so good in most people's opinion, but I've got to tell you, I love them. They save a ton of work. Give me Roundup Ready every time." Runoff from those fields sometimes poisons local water systems and neighbors' wells. Seven years ago the next town down the river warned its residents not to use the city water because runoff from heavy rains on the farm hills had polluted it to dangerous levels.

Our town is growing, not exponentially, or even rapidly as some places are, but growing. The loss of agricultural land is an issue. Some don't see it as a critical concern, because as a nation we still think we grow more food than we can use. What does the loss of a few more farms, a few more acres, matter if the slack can be made up with increased production by larger farms? Perhaps not much. But we do worry about metropolitan sprawl from the great cities not far from us. We like our town as it is, and we don't want to see it swallowed up by an advancing megalopolis. Our rural neighbors, on the other hand, worry about our own "mini-metro" sprawl. Folks outside our town know that our steep blufflands limit our building sites, and they worry about our spillover into agricultural land and open space. This is all the same subject, for one way to keep metro sprawl from engulfing us all is to help those folks who want to stay on the farm do so, making many small farms viable for people who love the land, and thus keeping small towns and schools and businesses viable so it is not necessary for us, or for our children, to flood the metropolitan areas and strain every urban resource. Until we make small farms and small towns viable again, metro sprawl will remain a threat for urban communities and will perpetuate the low-quality lives of the nongentry in the heart of major downtown areas. Keeping farms small and in the hands of many families works not only for the small farmer but for his urban cousins too.

And if those small farms follow sustainable farming practices and minimize the use of chemicals and energy and reduce toxic inputs into the soil, and if we could slow the energy waste of urban centers too, well . . . maybe . . . we could even develop a sustainable energy program. Would we therefore have a sustainable culture? Probably not.

Unless, of course, we could also develop our own human capacities for economy, for grace, for doing with less, reducing our desire to acquire goods and burn energy, and learn to turn our fear of others into more positive avenues like respect or even appreciation. For at the heart of all our present systems, our own human selves are the real issue. To create a sustainable culture that will support a sustainable agriculture and healthy soil, sustainable yields of fish or logs, sustainable water supplies or energy or clean air, to establish social justice in our own small towns and big cities so that the economy is viable for all and so that prejudice no longer figures in our reckonings about whom to include and whom to exclude from social intercourse, we have to work on ourselves. Everything goes back to that.

The arena for creating a sustainable culture is not really very large; the area we have to work in is about the size of ourselves, our friends, and a few broader relationships. What we discover when we work on ourselves is that our relationships cover a great deal of terrain, more than we ever thought, some of which we did not recognize before: our relationship to the rest of the natural world, a relationship that is critical even if we never leave our closest urban neighborhood, never venture out into the wilderness of Alaska or walk the Appalachian Trail, even if we never picnic in the most domesticated park in our local community or take a walk along the reservoir in the evening. We live in the heart of Nature in a city apartment as well as in a wilderness landscape. And if we live in a gated community shared only by others of similar income, ethnic background, and educational level, we are still part of the whole mix of the world's makeup. We can no more escape the society as a whole with its myriad shades of color, range of genders and sexual orientations, and multitude of languages, ceremonies, music, and dances than we can extricate ourselves from the Nature in which we are embedded and upon which we depend for our very lives.

It is possible to imagine a safer, more sustainable world. Not too far-fetched even. For all these thoughts provide us some clues to the nature of a sustainable culture, a future our grandchildren and their kids can relax in. There are no easy solutions, no quick schemes that will fix everything, and no blueprints or outlines that you can take to the classroom or office and make their transformation into sustainable institutions simple. But there are some ways to develop a worldview that leads to a brighter future, not just for us humans but for all creatures and for the great landscapes that we love. It's a thorny path, and steep, for a while yet, "the steep climb / of everything going up, / up," as Gary Snyder describes it.[3] But we do have some signposts to guide us. They do not all come from Western culture and traditions, nor from science and technology, for which Western culture is renowned. Some of them do, of course, for its oldest traditions are as noble as any culture's. But to limit ourselves to one worldview is to live with blinders. We need to see the world through as many lenses as we can find, adopting those worldviews that enlarge our grasp of our own capacities and strengths and that show us a way through this thicket that we are in.

There is enough evidence to indicate the kinds of environmental troubles we face. Articles that tell us how dismal our future is inundate us. I will not add long lists of expiring species, toppling rainforests, increasing ultraviolet rays, soil depletion, aquifer drawdowns that exceed the rate of their recovery, and other environmental damages that have become a litany in our common discourse. Instead I will try to think somewhat differently about how we see the world and what sustainability may be about.

Partly because of the complexity of the issues and their maze of interrelationships, no one seems able or willing to define sustainability.

This book is predicated on the notion that sustainability is not only about the environment or about the economy or about social issues and social justice or about a kind of education that would allow future generations to save themselves. It is about all four—and more. Our current dilemma is that we have to work on all these fronts at once, without a clear definition of what sustainability is. That is a bit like trying to build a house without a blueprint. On the other hand, trying to pin down a definition of sustainability is about like tracking bees through a blizzard. This may be in part because our definitions often come from single disciplines and are simply not comprehensive enough. And that may be because the disciplines don't talk much to those practicing sustainability in other arenas.

Twenty years ago the World Commission on Environment and Development spoke of a "sustainable development" that doesn't diminish the opportunity for future generations to have a decent life.[4] It was an unfortunate choice of words. Since then, in the field of economics we have talked about sustainable development as if it were a synonym for growth. The emphasis remains on development rather than sustainability. We still equate a healthy economy with gross national product or with consumer spending. The greater the spending, the greater our citizens' faith in the economy, we say. Yet the rampant consumerism in our country is an indication of disease rather than health, and it is one indication that we do not have a sustainable economy or a sustainable culture. Even sustainable agriculture often tends to focus on increasing yield: the number of hogs raised or bushels harvested. But a sustainable agriculture clearly means something more than higher production.

Neither is sheer longevity an adequate definition of sustainability. One often hears references to a "Native American view" that we should consider the impact of our actions "out to the seventh generation." Yet none of us want to just keep on getting older and older forever. Species, in the natural course of things, come and go. None last forever, except maybe cockroaches and sharks, and I suspect that what enables their longevity is not their strength and fitness but their hiddenness, their dwelling in the dark and the depths, their innate shyness. We might learn something from them, too, given our American proclivity for display, for putting ourselves forward, flexing muscles, telling the world a story that we are tough, man, tough; don't mess with us. Sharks and cockroaches, on the other hand, do not display; they simply do and then disappear, twisting into the deeps or scuttling into the dark. They never care to attract attention to themselves. Sustainability is about more than exhibitions of power and about more than going on forever, though longevity is involved as a secondary characteristic.

Rather, sustainability has to do with living with changes in the earth, learning to live with ebbs and flows and seasons, the comings and goings of our fellow humans and other species. Environmentalist friends talk about learning to live with Nature. I have some reservations about that, which will show up later, but maybe it will work if longevity is coupled with a sense of rhythm, an appropriateness of scale—all the old formulations of E. F. Schumacher. We will work our way toward a definition of sustainability by looking first at some critical issues the culture has to face before it can become a culture that can at least survive.

Whether we have a definition or not, we do have some models of sustainable cultures ready at hand. Like Snyder's "Axe Handles," they offer us clues to "how we go on." What is clear is that the issues are so complex and interrelated that no one discipline can resolve them. It will take all our scholarly disciplines, and all our personal self-discipline, to create a sustainable culture. Indeed, we have to do more than slow down or stop the loss of diversity; we have to regenerate diversity. Perhaps what we need to work toward then is the regenerative culture that Ben Webb speaks of—a concept that fits our task in both the environment and our social and personal or spiritual lives.

Thinking about this in terms of our education systems, a regenerative culture uses literature, history, the sciences, the arts, and philosophy as vehicles for recreating—regenerating—itself. Sustainability then becomes an umbrella under which every means of inquiry is brought to bear on helping us to learn how to live wisely on the earth, with each other, and within ourselves.

But sustainability is not only a metaphor, or an umbrella under which we can marshal all our resources. It is also about integrity—the integrating power of a comprehensive view of the world and all its complex systems. What needs regeneration or restoration is not just the culture or ourselves but the integrity of complex ecosystems including our human hearts. Such a view regenerates the integrity of our study, thought, and action so that the integrity of environmental systems is restored, our social systems' integrity is renewed, and our sense of our own integration, our own human integrity and dignity within the natural and social systems with which we are inextricably entwined, is sustained.

Sustainability and regeneration are thus also about wholeness. The integrity of the economy, our stories, music, science, the arts, and philosophy lies not in their profit status or their earning power or their bottom line but in their healing service to the life of the world. In a sustainable culture, whatever tends to dissolve integrity is seen as no longer profitable, for profitability is also part of the issue that concerns us when we consider sustainability. To increase our cultural profit margin, we may have to scale back, downsize, get rid of some luxuries, increase our participation in the community, develop mutual respect, and open youthful minds so that they are not susceptible to hustlers of any kind of snake oil, whether political, commercial, educational, or religious. Then they can create music, tell stories, write poems, and prepare themselves to collaborate in a world race toward sustainability.

Obviously these are just some of the characteristics required to create a sustainable culture and a sustainable life for all, but any conversation about sustainability must include at least these. And this word of caution: sustainability is not a state we achieve once and for all. It is a process of working toward, forever, being aware of the desired characteristics and alert to the implications of our choices every day. Does this story, song, idea, plan, policy, or poem move us toward sustainability? This is the question we ask ourselves, and we have to ask it every day for all time.

Defining Sustainability

THERE ARE CLUES ABOUT what sustainability might mean. As Ted Chamberlin suggested in a previous chapter, learning may be more about recognitions than about definitions or acquiring information. Perhaps we want to recognize rather than define sustainability. One way to create such recognition might be to ask, What are the characteristics of a sustainable culture? If we can determine those, then perhaps we can back into a recognition. Indeed, a sustainable culture begins in recognitions.

A sustainable culture recognizes relationships. That is, it knows that everything is connected. It knows that, as folks like John Muir, Wendell Berry, and Gary Snyder have said, You can't do any one thing. The farthest reach of the spider's gossamer web trembles at the slightest touch. Our whole system of environmental, social, and personal worlds of intelligence and spirituality is also gossamer, equally sensitive, equally fragile, equally interrelated. A sustainable culture recognizes that all health—human health, the health of other species, community health, economic health, and the health of our institutions—is related, and all health is directly tied to the health of the soil. None of the former is possible for long without the latter. A sustainable culture recognizes the relationships between humans and other creatures, from microviruses to watershed ecosystems to global ecology to the cosmos beyond our globe, and seeks to create healthy relationships with all.

A sustainable culture takes care not only of the human species but others as well, protecting biological diversity. A sustainable culture also takes care of ethnic diversity, including language diversity, ceremonial and ritual diversity, and diversity of worldview, recognizing that these diversities are as critical as biological diversity to the world's survival. "Men are like plants," J. Hector St. John de Crèvecoeur wrote in 1782; "the goodness and flavour of the fruit proceeds from the peculiar soil and exposition in which they grow."[1] The differences between us are as fruitful to the whole as the commonalities.

A sustainable culture is a socially and economically just culture, recognizing that wealth is never an individual or corporate accomplishment. It is a gift of society. That may rankle, since most prosperous folks see the acquisition of wealth as a personal triumph of persistence, hard work, and better brains than other folks have. Yet the corporation itself is a creation not only of individuals or partners but of the society that gives the corporation its charter and allows it to conduct business in the society's interest. Theoretically the society can withdraw the charter that allows its existence. The perpetual demands by corporations that Congress (and therefore society) make allowances for the free exercise of their economic muscle attest to the fact that, at some basic level, society grants us the privilege of generating our wealth. The perpetual presence of transnational attorneys in the congressional committee rooms during markup and when regulations are drawn up following a bill's passage also points to society's role in granting us our wealth. Even the smallest commercial enterprises must have a license from society to conduct their business, and society presumes that business will be to its benefit as well as the proprietor's. Corporations and commerce therefore owe society some return on society's investment in their corporate success beyond the tax loopholes they manage to find and the grants and scholarships they give for public relations purposes. Indigenous societies recognize that such reciprocity, returning a portion of what has been taken, is a fundamental principle of sustainability. Mainstream American culture might appropriate such wisdom and then pass it on to the next generation as a wise gesture toward maintaining the culture. Following the path of reciprocity offers every citizen an opportunity to return something of value to the culture and to the earth in exchange for his or her own existence. Any system that deprives people of an opportunity to see themselves as useful members of the human community or that denies people a means of giving themselves to a larger community beyond themselves—including the earth community—is not sustainable. It will freeze to death of its own coldness.

There is no sustainable culture without a healthy intellectual and spiritual life. A sustainable culture incorporates systems for healing, reconciliation, and the return of exiles and affords healthy, nonviolent means of resolving differences. It provides all citizens a sense that we are in this together. A sustainable culture nourishes its artistic enterprises, recognizing that no culture can sustain itself for long without the arts, and if it could, it would not be worthy of the name. A sustainable culture also nourishes its academic efforts, recognizing that new knowledge about the past and the present in the sciences, the humanities, and the arts is essential to creating a sustainable culture. It supports its education systems for everyone, cradle to grave, recognizing that the young must be raised to understand and appreciate the values of the culture and to have the wisdom to grow beyond them. Helping children understand the culture and discover means to transcend the old ways is essential to keeping a culture alive, flexible, and sustainable. A sustainable culture affords its citizens avenues of meaning. What good does it do us to become wealthy if we cannot find any meaning for our lives? A sustainable culture understands the implications of its stories and tells the healthy stories over and over in proper season. In every culture we become the stories we tell ourselves; they are self-fulfilling prophecies. If the stories we tell about ourselves and about our culture's future are healthy, we will become healthy—and that health includes all the elements we have talked about thus far.

A sustainable culture has sustainable food sources and experiences food security. It also has sustainable energy sources. A sustainable culture has avenues of new development, including intellectual growth and spiritual development, not just economic growth.

In light of these recognitions, we are ready to look at a tentative definition of sustainability: Sustainability is a worldview. That worldview recognizes the relationships, the connections, that ramify through every aspect of Nature, including the human, and knows itself to be dependent upon the land.

Out of that worldview comes a set of principles for our conduct. Cultures that hold a worldview based on sustainability strive for, and achieve with varying degrees of success, a clear measure of balance and harmony in their societies and offer means for individuals to achieve balance and harmony in themselves. The culture's relationship to each element of the natural world's ten thousand beings is regarded as essential and given due respect. A sustainable culture honors that worldview and lives it out to the best of its ability hour by hour and day by day.

Out of these general principles come specific actions derived from direct observation and experience. Reciprocity, which is built into the system, can become an asset or a liability depending on people's attitudes and actions. If you don't have the worldview, you don't get the sustainability. Among the outcomes of a sustainable worldview are longevity, respect for self and others, and a satisfying and meaningful life in community with others.

Larry Gates, a fisheries biologist with encyclopedic knowledge of much else besides, told me, "We don't need a definition of sustainability; sustainability operates out of a system of principles." Yet I've tried to work toward a definition of sustainability in part because the usual view is that sustainability is synonymous with longevity, and I think that answer doesn't look quite hard enough at the concept. Longevity is the result of a process. The cultures that have lasted longest have lasted not because they are sustainable cultures, as if that were an inherent characteristic or a state they had achieved. Rather, they created a process that grew out of their worldview. The choices they made created the conditions for longevity, every day. Inevitably, those choices were dictated by their particular worldview. There has never been a sustainable culture without need to worry any longer about the future. Longevity is, has always been, a by-product of the culture's life, developing from the myriad choices about how and what to eat, how to dress, how to create shelter, how to maintain resources, how to treat our neighbors both near and distant, and how, finally, to treat ourselves—all with respect. The definition of sustainability lies in the worldview and ways of the culture, not in the outcomes we hope for.

Further, sustainability is not a state we reach but something we work toward forever. In that sense of it Larry is also right. What is great about cultures that have been around forever is not that they are sustainable but that they have lived, and continue to live, a sustainable life, making those daily choices that lead toward sustainability. Their success is in small part an accident; in far larger measure it is by design. They have respect for the world that supports them, and that has resulted, so far, in their longevity. We know that now because we can look back through history and see it, but every month, every week, every day of that long history, even in times of abundance, the culture was in jeopardy, teetering in the balance, its future seen but dimly, and then was shaped deliberately, with respect and hope, to keep it on a path that could sustain it.

Does that mean that life was perpetually stressed, anxious, and filled with drudgery? No. Occasionally, yes, of course, but not continuously. People learned to trust their observations and their actions, and they learned over and over, as their environment evolved around them, what it took to live in harmony with all creatures and the land. People also had confidence that if they followed the culture's wisdom, modifying it based on their own, more recent observations, then they would survive. Indeed, they followed the principles of sustainability, as Larry noted, but the principles lay in the worldview—that we must carefully and respectfully observe the world around us. That principle in the worldview informs the good practice that grows from it as the culture learns to express the worldview. There were times enough of plenty, and celebrations to honor those. Times of want were reduced as a new culture found its way and developed its wisdom with regard to the land and its food sources and as it developed rituals and practices to ameliorate times of hunger and need.

Sustainability for the oldest surviving cultures was never a present reality, settled, fully grasped and realized, assured. It was, and remains in our own culture, always contingent. It remains to be seen in a future we can only dimly discern. The sustainability worldview sees that it is better to live in a mutually respectful world than one filled with greed, anger, antagonism, hatred, and indifference that stem from inappropriate actions. Eskimo wisdom tells us we cannot afford to have the salmon turn their backs on us; therefore we must treat the salmon with respect. That respectful attitude toward the world starts with us, and it has to be nourished by our own interior mind and spirit.

Whether we want to define sustainability or not, it is hard to argue with the second part of Larry's thesis, that sustainability operates from a system of principles. One can quickly see that a society that gives precedence to cash, credit, or capital economics, or believes in economic determinism, is bound to be out of balance, lacking harmony, facing a short-term future. A culture that elevates humans to a status greater than the ecosystems that form our habitat, that sees the earth only as a means to personal or corporate wealth, is out of whack and has a foreshortened future. And any culture that elevates the environment to a status that prohibits humans from participation, as some wilderness advocates now hold, is also doomed to a short life.

The first principle of a sustainable culture is self-cultivation. I do not mean the self-fulfillment that contemporary Americans seem so desperate for but cultivating the self, not for the sake of the self but so that there is something in us for others. That "something" in us is the capacity for respect, for engagement, for paying attention. This is not the kind of respect that we want to grant every idea these days, as if the notions of equality and respect have to mean that every idea, wise or foolish, has equal weight or merit. Rather, the respect that is an operating principle of a sustainable culture includes genuine respect for every essential aspect of every relationship in our human culture, and those relationships clearly include our connections to ecosystems, other critters, and the soil. We are not born with a facility for respect. A capacity for respect may be innate, but to become fruitful it has to be cultivated by each of us. Such capacities grow through reflection, thoughtfulness, and a desire for wisdom, and when properly nourished they become the pathway toward the elimination of prejudice and poverty, toward the implementation of social justice and care for the environment, all essential ingredients of a sustainable culture.

Walter Rosenberry, a Denver teacher and friend, talked to me once about Samuel Sewall, a Puritan who fought against slavery. Sewall wrote in 1700 that black people should "be treated with a respect agreeable."[2] He meant, of course, a respect agreeable to blacks as well as to whites. Like Walter, I love that phrase. It fits so much more than just our attitude toward other races and cultures. It's appropriate for all of us who love our homeland and want to see it persist in spite of our numerous assaults upon it. It's appropriate, too, for those of us who love our culture's ideals, if not always its practices. To accord our land and all its creatures a respect agreeable; to accord our rivers a respect agreeable; to accord our communities, whether small or large, sprawling, and urban, and all their complex components a respect agreeable; to accord our small, local economies a respect agreeable; to accord our everyday discourse, our public and political rhetoric a respect agreeable—those are tasks worthy of the best in all of us.

The person who first drove this point relentlessly home for twentieth-century culture was Albert Schweitzer. He took respect a step further. He wanted us to inculcate "reverence for life" as both a personal and a cultural goal. He included in that reverence the lowliest and most obnoxious creatures. No matter our place in the food chain, one characteristic of every creature is the will to live, Schweitzer declared. That is our common heritage, our link to every other animal and to plants as well. That will to live deserves not only respect but reverence for the most fundamental force that links every creature in the world.[3]

Another principle from which sustainability operates is attention, our paying attention to the world, being wholly present to it all. This is a principle that ramifies through every aspect of a sustainable culture. I remember a cattleman in Colorado talking about how he knew when to rotate cattle from one paddock to another so that the grass could come back. "I spend a lot of time down on my knees, looking at the grass," he said, squinting up at me from his position on those knees. "You have to pay attention all the time." That is the kind of attentive respect that allows the grazing animal, the plant life, and the humans who depend on it to thrive. I told him that poets say paying attention is where poems come from. He grinned and said, "I'll be damned." Then he added, "Sounds right to me."

Martin Buber, the Jewish mystic and philosopher, saw being fully present as one of the essential ways that humans have of paying respect, of treating each other as "Thou," with a capital *T*, rather than "it," a thing or object: neutral, useable, demeaned, and ultimately disposable. We show our respect for another by being engaged, fully present with the other. Buber asks us to extend that courtesy to nonhuman creatures as well.[4]

Working from these principles in our own lives, and getting the culture to operate from these principles, is a task that will lead to a new story about a sustainable culture, a story about respect that all creatures, including us, can live with. When we understand that story and teach it to our children, we will have learned enough to respect ourselves as well as our culture, to respect the myriad things around us, and to respect both the present and the future. Then we may find ourselves at home, in our place, at last.

Stories for Sustainability

ALASKAN ORAL HISTORIANS Phyllis Morrow and William Schneider admit "we face an awesome responsibility: the certain knowledge that our words will return."[1] The responsibility rests not only upon oral historians but on all of us. Our modern words are often ancient, ricocheting down the years to us battered and banged around by time, misspoken till they are hardly recognizable echoes of other tongues, not only Latin, Greek, Navajo, Sami, or German but Sanskrit, Etruscan, or perhaps languages lost now without even a linguistic trace. What is lost is a matter not only of vocabulary but also of the way words are used to make up the great stories of any cultural tradition.

Our stories are the expression of our worldview. Perhaps even more than the principles from which sustainability operates, our stories are the primary force that enables cultures to become sustainable. Whether cultures or individuals, we become the stories we tell ourselves and our communities.

There is an old question that has worried philosophers for thousands of years: What is the difference between humans and animals? Some folks hold that we have intelligence and animals don't. But no other animals have threatened the life of the whole planet, and we humans certainly have, so who's intelligent? An old American Indian poem goes like this:

> The frog does not
> drink up the pond
> in which it lives.

Now we have growing evidence that the Indians were right and that frogs may be smarter than we are. Other folks hold that we have language and animals don't, but chimps and whales and chickadees, coyotes and treehoppers, seem to communicate with one another very nicely, while I often garble my words.

So where does the difference really lie? We have to start by admitting that we are animals too. But each species has some characteristics that set it apart from all the rest. For humans one of those characteristics is that we tell stories, and so far as we can tell, other creatures don't. We may yet learn that animals tell stories too, but for now it seems fair to say that one distinguishing characteristic is that humans are storytelling creatures. We can't help ourselves. We go out in the morning, and when we come back, our spouse asks, "How'd it go today?" and immediately we are telling stories.

The great stories last and return to create us. They go out and come back. The past, put before us in a story that is healthy, is at the same time put behind us, and we are healed. The healthy story is our refuge; its respite enables a new freedom in us: past and present become commingled, and the outlines of the future, though vague, are made visible, for the seeds of the past, we may be sure, are growing into the present and will ripen, fall into history, and grow again in the days to come.

"A story does not exist to be captured," say Morrow and Schneider, "but passed on."[2] The stories I have in mind are both oral and written, with a slight tilt toward the oral, for ours is still a far more oral culture than we like to admit. In the cloakrooms and halls of our legislatures, our government representatives talk over legislation. In conversation they make agreements and cut deals. Often when they part, one says, "Get me a memo on that, will you?" and the other agrees. Both know the memo is not for reading; it's for the file, in case some backup is needed later. Literacy is almost irrelevant to the conduct of such business. Legislators often do not read a bill under debate but ask a staffer to tell them its salient points. On the personal level, we speak of friends, of those we love, through stories—nearly all of them oral. The healing times in our grief come through the stories we tell about a loved one's goofiness, the "Remember the time" stories when he or she said something funny or pulled off a good joke, or the more serious stories about the good work they performed on a difficult task in the community. And how do we keep them before us after they are gone, except through stories?

I knew a schoolteacher in Red Lodge, Montana, whose name was Estelle Province; she taught in Red Lodge for nearly forty years and then retired. When she was eighty-one, Stelle decided to take a trip around the world. "I've always wanted to do that," she said, "and it's about time," so she took off on a group tour. I got a postcard from her marked Hong Kong. All it said was that she was leaving the tour and going to travel on her own. When she got back, I said, "Stelle, what happened?" and she replied, "Oh, you know that tour—it was just full of old people, and they weren't any fun." So she finished her trip through India, the Middle East, across Europe and England, and back across the United States by herself. One time, coming out of church on Sunday morning, Stelle leaned close and whispered, "You know, I'm reading one of those *modern* novels, and it's just *so* vulgar—I can't put it down." Stelle's gone now, but people die by degrees, little by little, as we stop telling our stories about them. The stories about our friends and family members get farther and farther apart and then fade. When the stories are gone, they are gone. As long as I can tell Stelle Province stories it's almost as if Stelle were still around, providing good companionship, teaching me how to grow old more gracefully than I am really inclined to, teaching me how to live. When that happens, the story becomes the medium for life itself.

So the stories that may be most important are not the great works of literature like *Moby-Dick* or *Macbeth* or the *Iliad*. Those stories are important for reflection, meditation, wonder, thinking through our own experiences when we are older, but they come along too late in our lives to be really formative. The most important stories are those we hear from our elders when we are young: the kinds of stories that boys in my generation used to hear in the living room after Thanksgiving dinner when the men gathered to talk, or the stories that girls heard in the kitchen after Thanksgiving dinner when the women gathered to

work. Those are the stories that tell us what our elders admire and respect and what they disdain or despise. It's those types of stories, many
indigenous people say, that teach us how to be human. If those stories
are as valuable as I think and as critical as the Koyukon, for example,
believe, then a culture is in serious trouble if its kids grow up without
hearing healthy stories from their elders, especially if the stories that
take their place come from television and film. Those media stories are
not, as critics often claim, mindless. More often they are very mindful,
deliberate, and filled with purpose. But their purposes are so shallow
(corporate profit, entertainment) and so impossible to verify in our
own experience that they subvert the healthy stories that might lead to
a sustainable culture.

But like every other subsistence culture, we have to not only hear
those stories from our elders but also experience their truth for ourselves, to test them in our own experience. As a child, I heard a story
about a neighbor who raised sheep. He'd been knocked down by one of
his rams. I did not understand the import of that story, or perhaps I did
understand but ignored it. I was about five years old and clearly thought
I was smarter than stories. I went into the corral behind the barn on my
grandfather's farm, crawling through the rough boards that made the
fence. There was one sheep in there, a ram. Two horses were there too,
standing hipshot across the corral, and I wanted to get to them—
another action I had no business undertaking. I seem to remember
Grandpa telling me not to go in there. Of course I went anyway, could
hardly wait till no one was around. When I had my back turned, walking toward the horses, that ram charged and hit me so hard in the butt
that I flew through the air and landed in a heap against the fence, unable to breathe. I learned some things from that experience that I had
heard but not learned from Grandpa. One thing was about Grandpa's
wisdom. Another was about the nature of surprise.

Because I was totally surprised, I was angry at that ram. The ram
was only doing what rams do. I suppose I was only doing what kids do.
But I was mad at that ram. Nowadays, when folks say that bureaucrats
hate surprises, that they are not flexible and that is why bureaucracies

are so slow to respond to human needs, I think about that ram. I've been a bureaucrat, and what folks say may be true, but because of what happened to me personally when I was five, it's occurred to me since that there aren't many farmers who like surprises either. And I have never met a pilot who appreciates surprises, like a sudden shift in the weather or a sudden loss of power. Closing on a house we've purchased, neither we nor our real estate agents like surprises, and neither do attorneys in court. Perhaps surprise means an unpleasantness more often than it does a treat, and most of us do not like surprises even when they come along in the natural course of events. Learning about sheep in my own experience verified the story I'd heard about sheep and installed it permanently on my brain's hard drive.

I remember another scene from my grandfather's farm. Still about five or six, I was standing outside the pen where Grandpa kept his pigs. At my feet was a galvanized bushel basket Grandpa used to carry corn to feed the hogs. There was one ear left in the bottom. I reached down and threw it into the pen. Those hogs converged on that corn, screaming like humans in great pain. I had never heard pigs make such sounds, and I was terrified, not knowing what to do to quiet things down. Grandpa came running out from the barn to see what had happened and quickly sized up the situation. He was a gentle man who would never dream of hitting or spanking me, but he said something in a tone of voice I never forgot: "Don't *ever* throw one ear of corn in a pen full of hogs," he said. The memory of those events, turned into a story I can tell others, shapes my view of the present. Now, every time the legislature meets to quarrel over tight budgets, I think about what Grandpa told me, and it informs my political views.

Elias Canetti, winner of the Nobel Prize in Literature in 1980, sees the role of the writer in the culture as that of *dichter*. Dichter can be translated as "speaker," "teller," or "talker," but in *The Conscience of Words* Canetti seems to connect with indigenous peoples in seeing the writer as a storyteller who gives his people "true stories" that "help make us human," as Koyukon linguist Eliza Jones describes her own people's sense of stories. Like Confucius, Canetti would rectify the names, but he adds an extra margin of responsibility to his definition of the storyteller's role: as someone who lives by words, his responsibility is "to stand against his time," for it is a time "which mollycoddles death." The way we save the world in our time is to stand for a sustainable culture, for such a culture always stands against death. Words are one powerful means of fulfilling that responsibility. It doesn't seem possible that mere words could have such power, but Canetti points out that "words, deliberate and used over and over again, misused words, led to this situation in which war became inevitable. If words can do so much—why can words not hinder it?"[3]

But why should Canetti's commitment to stories be limited to writers? The writer, the teacher, the politician, the public administrator, the scientist, the clergy member, the agriculturalist, and the lawn care specialist all have one common task among their other chores: each of us is called to be the dichter, the teller of stories whose language is always true, the teller of stories that will show us how to be human. This is step one in our creation of a sustainable culture. The economy has come to stand for death; military might finds it merely a means to an end. To adapt Canetti's notion, giving it a mildly Confucian twist, we are called to create and tell a story that will transform ourselves and the culture, a story that will sustain the world.

If Freud taught us anything in the century just passed, it is that there are some stories so painful that we cannot bear to tell them, that some stories can be buried so deeply inside us that we cannot even admit that we know them. But if we can call such painful events to mind, turn them into stories, and relate them to others, then a kind of healing can begin.

In 1964 I met and hunted caribou with Guy Groat, a bush pilot from South Naknek, Alaska, who became a friend. Hunting was but one kind of adventure with Guy. Flying anywhere was another. Guy was noted for getting to places that others wouldn't try, and occasionally I would hear him glide over our house in Naknek, making his final approach to his home strip in a heavy fog. On such days there was never any question but that it was Guy, because no one else would be up in that weather. No matter how good a pilot you are, though, the odds always catch up with you. In January 1965 Guy took off from Kodiak Island in clear weather to fly back to South Naknek. The weather closed in, and he couldn't land on his home strip. He radioed his wife that he was headed for King Salmon, but everything was socked in, a whiteout, in King Salmon, where even a U.S. Air Force station with its navigation aids could not guide him to a safe landing. He told King Salmon radio that he was heading back to Kodiak. That was the last we knew for ten days. The whiteout never lifted in all that time, and no one could get up to look for him. When it finally cleared, Fred Cunningham and I rode as spotters in Georgie Tibbetts's little Piper Tri-Pacer. We flew long rectangles three hundred feet over the tundra while other planes did the same, all of us flying adjacent patterns that would cover a wide swath of the Alaska Peninsula between King Salmon and Kodiak, making allowances for Guy's being blown off course in the foul weather. We finally found the wreckage on a ridge above Becharof Lake. We could see the scattered parts of the plane, partly drifted over by the snow, Guy still strapped in the cockpit where he had died instantly, snow drifting in around him too, where the door had been torn open when a wing came off on impact.

Freud had something in common with English poet William Wordsworth, who believed that the origin of poetry lies "in strong emotion recalled in tranquillity." One needs time to gain perspective, Wordsworth believed; then when one goes back to an incident to write about it, the emotion returns with its original, powerful creative force.[4] It was ten years before I got around to writing a little poem about Guy, and I did not even know I was thinking about him. I had buried his story somewhere down inside me. Freud's notion was borne out. In drawing that event up and turning it into a story, I experienced a kind of healing. Wordsworth was right too. If I had tried to write the poem earlier, the material would still have been in control of me. Time had given me enough distance that I could make it a story, but once I got started, the old emotion, the sense of loss and distance—that uncoverable distance between my safe place in the air and his broken body on the ground—came back to inform the poem. When that happens, friendship and loss become something more than strong emotion; they increase respect for life and loss, and both take on a sacred aspect. Now, as with Stelle Province, telling his story keeps Guy present in my life, and sometimes even in the lives of others who never knew him. The story becomes the medium for life itself; it sustains me and adds a dimension to my world. Such stories teach that to mollycoddle death is one of the ultimate sacrileges.

But we have to be careful about the stories we select and construct, for we are more apt to remember our stories than the actual events. That's just fine, but it does mean that we must choose our stories with some concern both for their truth and for the kinds of persons we want to become. If we rely on a history that is false, then the stories we tell ourselves about who we are will also be false. When that happens, we cannot establish integrity within ourselves or between ourselves and others.

We can be sure that we become the stories we tell ourselves about ourselves. If I am looking into the mirror in the morning and thinking about the office and telling myself a story about being the meanest and toughest guy there, then that's what I'm apt to be, and my anger will bite into my soul and warp my life. If I tell myself a different story in the morning, I'm apt to be a different person. Not immediately, of course, for the old stories we remember feed into us as powerfully as the newest stories. How sad, then, if the stories we remember are filled with absences or emptiness, with resentment, anger, or hate. Or if we spend so much time glued to the television that we never hear any stories from our elders. Or if our elders tell us stories that are untrue or full of anger and the desire for revenge. Then the stories we grow into become a trap, another kind of sickness rather than a means to health.

In *The Haunted Land: Facing Europe's Ghosts after Communism*, Tina Rosenberg quotes Karl Marx: "Men make their own history, but they do not make it just as they please; they do not make it under circumstances chosen by themselves, but under circumstances directly encountered, given and transmitted from the past." Rosenberg, with considerable wisdom, continues, "The gifts of memory and tradition are among humankind's greatest blessings. Many oppressed peoples can thank the weight of tradition for their very survival. But they can also thank it for their continued oppression." How can that be? "For too many governments, dealing with past injustice has been not a way to break free of it, but the first step in its recurrence."[5] There is a tie here with child abuse, which is another kind of oppression, repression, and violence. Abused children often grow up swearing that if they have children they will never treat them the way they were treated by their parents. Yet abused children are among those most apt to abuse their own children, reliving the pain they suffered by inflicting it on their own youngsters.

As with individuals, so with cultures. One can extend the notion of story from the intensely personal to the culturally essential, for history is one critical ingredient of a sustainable culture. It is the story of our common past, a way of remembering for a whole culture. If the culture grows up remembering slights—or worse, murder or other atrocities— perpetrated on parents and grandparents and beyond, then its stories inevitably breed warfare, and pain multiplies on every side. I'm not suggesting that we gloss over the evil that exists in the world or the violence that we have perpetrated and endured; one wants to live in a world that is real. We remember the evil and the painful not so that we can grow into them but so that we can muster the courage to confront evil, surmount pain, and, as Canetti insists, stand against death. As individuals and as citizens, then, we remember as much as we can of our stories, both the good and the painful, because without a story the present is less sensible and less bearable, the future more difficult to discern and more frightening. The great stories of any people nourish us, reveal us to ourselves, inspire us with the deeds of heroes, and bring us to the edge of despair.

Not only history but the humanities as a whole are primarily about stories and about the impact of our stories on our personal and cultural lives. When the humanities—history, literature, philosophy, criticism of the arts, and the social sciences—are ignored or forgotten in our public life, we handicap the humanities and impoverish our public life. The humanities then retreat to studies that are esoteric or personal, even trivial. But for our civic life, the loss of a humanities perspective is of even graver consequence. Without the humanities we pursue policy with no sense of history, no sense of what has been tried or how it worked out. Without the humanities we develop policy unaware of philosophy's story of the necessity for internal coherence and consistency with other policies. Without the humanities policy is determined outside the references of literature, and with no recognition of the options and opportunities offered to us by our study of languages, the source and the vehicle for all our stories. Without the humanities we forfeit the opportunity for discussion of the ramifications of our actions and abandon policy to whim, caprice, the computer model, or some other authority.

It is hard to bring the humanities to bear upon our public life. When we try to make our long history of stories, and the new ones we create every day, relevant to life in the world, we always feel, to quote T. S. Eliot, that we are making "a raid on the inarticulate / With shabby equipment always deteriorating / . . . under conditions / That seem unpropitious."[6] Powerful voices in our time raise lucid arguments against the prospects for success when the humanities wrestle with public policy. The naiveté of humanists working away in their ivory towers, the abstractness that seems unrelated to realpolitik, the "watering down" of the humanities when we try to make them work for us on public issues, all are called up as justifications for keeping policy and the humanities at a distance from each other. Oscar Handlin, writing in *Truth in History*, appears to argue that it is both dangerous and disruptive to history's integrity even to try. "The historian's vocation depends on this minimal operational article of faith: truth is absolute; it is as absolute as the world is real. . . . Truth is knowable and will out if earnestly pursued; and science is the procedure, or set of procedures for appropriating it."[7] Handlin is but one among many voices—William Bennett is another—apparently sage, but in reality sophist in their arguments.

But I am not dissuaded from my belief in the power of the humanities to affect public life, for there are other voices, far older, wiser, and more experienced than Bennett and Handlin. Livy, one of the early Roman historians, addresses us with utmost clarity about the "faithful effect" of history, presenting "real examples embodied in the most conspicuous form: from these you can take, both for yourself and the State, ideals at which to aim, you can learn also what to avoid." And over the dissent of Handlin and Bennett, Cicero says, "Our Senate always identified advantage with principle." Down the years those endless Roman senatorial arguments still echo, exploring the varying attributes and guises of advantage and principle from every angle. "This is the type of a problem," Cicero says, and he tells us a story: "Suppose there is a food shortage and famine at Rhodes, and the price of corn is high." When the first seaman lands with a hold filled with grain, should he charge whatever the market will bear, taking advantage of people's desperate hunger, refusing to sell to those who cannot afford his jacked-up price? Or should he tell them that just below the horizon there are other ships loaded with grain, and they will be in port by morning? Change corn to oil and his example could have come from yesterday's *Tribune* or *Times*—except that Cicero's question never even occurs to the Mobile executive, and if it did, and he asked it, he would be fired. "Why am I telling you these stories?" Cicero asks. "To show you that our ancestors would not countenance sharp practice." He begins another story, "Suppose that an honest man wants to sell a house because of certain defects of which he alone is aware." Does he reveal the defects to a prospective buyer? On and on the conversations range, over every aspect of personal and public life, business, politics, and international affairs. And every time the issues are highlighted by the humanities and clarified by stories. Cicero's stories from history and personal experience still create a path for us through the ethical wilderness that surrounds us.[8]

Koyukon Indians in Alaska tell different kinds of stories at different times of year, with different purposes in mind. "*Ka dont sid nee* are fall stories," Eliza Jones explained at the 1993 Sitka Summer Symposium. They are stories of how the world was formed and when animals lived as people. "There is no room for innovation in these stories," Eliza said; they are told the same way every time. But there are also *Yo'o gha dona* stories, stories from history about war, famine, hunting trips, grandparents. Unlike fall stories, these are personal stories, and they can be invented. "Sure, make up your own stories," Eliza says, knowing how stories shape us. "It's your life." *Toobaan atsah* are the stories for young people. An adult tells the story one night, and the child tells it back the next night. If the child says a word wrong, the adult says it correctly but without interrupting. The purpose of each type of story is clear: "When you tell a story, you are telling it for a good life or for good luck. You make the animals happy so they will give themselves to you." Those stories have at least one thing in common with many of our own stories from days past: "You should start off by saying, 'In the time very long ago,' and end, 'I thought the winter had just begun, and here I've chewed up part of it already.'" Other cultures also have differing stories for differing purposes and seasons.

If our purpose is to create a sustainable culture, then the most important stories will be stories about respect, about times people cared for one another or cared about the larger world. Those are stories we, too, tell for good luck or a good life. In our tradition many of those stories are simple, often told around the supper table, stories about everyday occurrences like one my dad told about Guy Briggs confronting an abusive customer at the post office where Pop and most of his friends worked. "He never lost his temper; he treated that woman with perfect respect clear through the whole thing. She was practically yelling in his face, but he never raised his voice. He was a real gentleman the whole time." Or the stories that come from the time a neighbor, a widow who lived alone, broke her hip, and she never had to cook a meal for herself till she was easily mobile again. Or the farmer whose barn burned and the neighbors not only came to help put the fire out before it spread to other buildings but also found time to rebuild the barn. Or the old mountain man who made an impression on his trapping partners by refusing to shoot the first buffalo that came along when they were hungry, waiting instead for a barren cow whose loss would not damage the herd. Or the cattleman whose rangeland conservation idea, learned from his cowboy father, was "Take half and leave half, and you'll always have plenty of grass." And he did.

Or stories about how communities are learning to work together on common problems. That is harder to pull off than one might think, but simple, homegrown stories can help us learn how to work together toward larger community goals. A few years ago, Robert Putnam argued that we as a nation were losing our social capital. He meant that we don't participate as much in public life as we used to.[9] That's a story I heard many versions of when I was traveling intensively around eight western states a few years back. What I haven't heard mentioned in expert testimony is an additional factor that stems in part, perhaps, from our falling away from public life: "We don't work together very well in this town" and its inevitable tagline, "There's a good ol' boys network that runs this place." Now that's a story I've been told in many places in the West and the Midwest. When one probes a little deeper into the life of our citizens or our communities, that story becomes a powerful theme: I don't have any control over what happens because somebody else runs things. As an Eskimo friend out on the Kuskokwim River once said, "All my life, I've been at the mercy of forces beyond my control."

Most often the good old boys run the town not out of some ulterior motive, or the desire for power, or because they are manipulative. We all have ulterior motives, even when bent on doing good, and many of us get manipulative when it seems that's the only way to make something happen at home or in the public arena—and unlike many of us, those "town fathers" already have the power. One reason *they* run the place is that not many other folks have either the inclination or the time to pitch in and help. Or, let's admit it, we feel that we don't have the prestige and authority. Two things are fundamental in this story about who runs our towns: One is that not many folks think that what those guys (and until very recently it has been all guys) do is bad. They've fixed up some things that needed fixing; they've brought some things to town that we all need. But the other thing is that they don't allow the rest of us in on the secret. We can't get involved because they do it over breakfast at the hotel or lunch at the café—both of which they happen to own. It's not the action but the process that rubs us the wrong way. But getting other folks involved in the process is hard, time-consuming work. When the good old boys want to get something done in town, they want to get it done as expeditiously as possible, with the least amount of effort on their part and the least amount of their time consumed. The time and inclination they don't have (since everyone's time and inclinations are limited) is to go recruit a bunch of other folks to help. It's unnecessary, for one thing, because they can get it done themselves. Their lack of time and inclination and ours often intersect and render our participation impossible.

But suppose, as happened in our town, some folks decide they really would like to get involved in thinking about the future of the community, and suppose those folks recognize that for a positive future to have a chance, all the stakeholders in town need to have a voice. In our town a few winters back, we agreed that there were some issues that we needed to address, and that we wanted to address them together. This was not a dissident group but ordinary folks whose purpose was to be as inclusive as possible—"town fathers" and "unusual voices" agreeing to work together on some scenarios, creating some plausible stories about our plausible futures.

We recruited thirty people from as many of those stakeholder categories as possible, and for the first time in our history we included people of color in the process, as well as single moms, commuters, regional farmers, small business owners, waiters, and clerks, as well as captains of our local industry. We put together a weekend workshop, directed by an outstanding international consultant named David Chrislip, to develop our scenarios. David had a little questionnaire, complete with an interview protocol. I interviewed more than half the participants in the workshop using that protocol and talked with dozens of other people in the community less formally. The purpose of the interviews was to uncover the deepest concerns of our citizens and chart some of the driving forces in the community that were not part of the general public discourse. This was important because our scenarios were not to be based on extensions of data such as those the old strategic planning model uses but were to be ways of revealing the underlying issues that we were really concerned about.

There was some apprehension among the folks who run things that the unusual voices we recruited couldn't know as much about the town as they did, and wouldn't have enough of the information they were privy to, to arrive at the threatening issues that the town fathers were aware of. That apprehension was not borne out.

David organized the responses to the workshop interview questions into clusters, similar concerns falling quite naturally together. Those concerns were presented to the members of the workshop. A long day of intense conversation and winnowing led to the selection of two issues of utmost concern. One of those was predictable: we were faced with a sharp loss of tax revenue caused by changes in the state tax structure for energy companies. That was precisely the issue most of our community leaders were focused on. The other issue was not predictable at all: "We don't work together very well." It took but a short time in our conversation to discover that this concern should have been predictable because it was almost universally shared, but we had never developed a means to bring such matters to the surface of our public conversation.

Our scenarios then developed along a double axis that led to four stories we could tell ourselves about our future. The first was our worst-case scenario: What if we never do learn to work together, and we get a big tax hit as the tax base falls? Second, What if we do learn to work together, and we get a big tax increase? Third, What if we never learn to work together, but the tax loss is moderate or smaller than we fear? The fourth was our best-case scenario: What if we do learn to work together, and we get only a moderate hit in our tax revenues? That was something to work toward! David's expertise in leading workshops let all the other concerns that were raised in the interviews—the loss of open space, the role of the arts and humanities in the community, the various concerns about industry and commerce and their impact on our economy, schooling, the environment, and many others—come back as part of the story lines that we developed about how the future might unfold for us.

This community effort, reported in a special insert in our local newspaper, did not give us a plan to follow—our city and county officials work on those—but it allowed us to prepare ourselves for a future that is largely unpredictable. We would check ourselves not against a plan but against our stories, trying to avoid some of the pitfalls that became clear as part of the discovery involved in creating the scenarios, and trying to make some of the good possibilities happen.

The workshop also showed us that we might learn to work together. We have not learned that yet, and one shot of collaboration in our collective arm will not make collaboration last, but it does make it seem possible, and desirable. The people of color who took part were recognized by others in the community in a new way and were soon invited to serve on boards and commissions in town—also a first for us. In creating stories we can tell ourselves about the future, we also created a new story we can tell ourselves about who we are now, a story about inclusiveness that creates a stronger sense of community. I do not mean to exaggerate this, for we have taken the tiniest baby steps toward creating a sustainable community, but we have taken a step that came out of stories we had to tell ourselves, and those narratives created more stories for us to tell.

Here is another story about how place becomes a healing story, recounted by Ted Bernard and Jora Young in *The Ecology of Hope*. It begins with a statement by Sanford Lowry, a rancher in the Mattole watershed on the northern coast of California: "When you get right down to it, everybody who lives here—the ranchers, the enviros, fishermen, the timber companies—are neighbors by virtue of sharing the same place. If we don't manage our own affairs, someone else will. As a landowner, I've got to be involved. I don't want to stand idly by. I want some say because I expect to keep my ranch intact and the land is important to me and to my children."[10]

The Mattole was historically the domain of Athapaskan-speaking Mattole and Sinkyone peoples. White settlers moved in during the 1850s, and the Indians were mostly wiped out in the 1864 Squaw Creek massacre. One report on the area noted, "In the span of eleven years, a culture and people which had been in place for hundreds or thousands of years was completely decimated." They and their language became extinct. The valley itself was akin to paradise—if one could stomach hard work. Salmon and steelhead ran in the "cold, stable, deeply channeled waterway [the

Mattole River] enclosed and cooled by riparian vegetation." There were crop and cattle agriculture with good harvests, and there were trees for the cutting. Over a thirty-year period, from 1950 to 1980, the hills were stripped and the loggers mostly moved on, leaving only a remnant to harvest the 9 percent of old growth that had not yet been cut. "What was left behind was an exposed steeply sloping landscape, subject to frequent tremors and quakes, in one of the wettest places in North America. . . . The soils quickly found their way to the Mattole, which turned into a shallow, braided stream with broad, cobbled floodplains, warm in summer, flashy in winter," write Bernard and Young. Salmon were choked out of the silted river, and between 1981 and 1991, the chinook salmon escapement upstream declined from 3,000 to approximately 150. The coho salmon escapement dropped from 500 to fewer than 100. That "got the attention" of the residents.[11]

From a chaos of conflicting and sometimes virulently antagonistic interests, mostly centered on the final cutting of old-growth forest in the late 1980s and early 1990s, there grew an agreement that something had to be done. Local citizens created a "restoration council" to work on the river. On the assumption that you can't fix a thing if you don't know where it's broken, a community-based initiative began to gather detailed information about their place, and citizens began to write their own geography. They studied their history, ecology, and geomorphology and paid attention to the cultural details. They plotted sites of erosion, developed topographic maps, and gathered aerial photographs, building a solid base of information out of which they could create a new story of their life in that place. Bernard and Young note that it was the local geography that reconnected citizens "to the ecological systems they inhabit and reinhabit."[12]

One member of the council said, "By spending the time to reorganize biotic, geologic, and demographic information into a watershed context, we are ritually reanimating a real place that had become totally abstracted." It was that abstractness that had caused the fragmentation and the clashes just a couple of years earlier. But now, he continued, "Our maps of salmonid habitat, of old-growth distribution, of timber harvest history and erosion sites, of rehabilitation work, our creek addresses for watershed residents, become, when distributed by mail to all inhabitants, the self-expression of a living place."[13]

As Sanford Lowry indicated, it was essential that all the stakeholders work together. It was not easy, and the authors' reports make a few public meetings sound raucous, painful, and discouraging. But some folks persisted, and they won others over. They have not yet saved the watershed or achieved full consensus, and none of it has come without overcoming friction—and not all the friction has been overcome. But the salmon runs are slowly coming up. It is too soon to tell whether the effect of the river restoration effort will sustain itself. The effort "sometimes resembles a car with 16 wheels, four of which are always flat. A different four on different occasions. It's a slow process," a council member noted. But it is a process. And so far it continues.[14]

Bernard and Young tell the story with obvious relish, for they are convinced that "this is about a different kind of resource management based neither on political constructs nor resource warfare, but rather on the way nature actually works." Another hopeful indication of future success is that "it centers around a unit of inordinate natural significance, the watershed, and on mutual concern for the health not only of this watershed but also of the human economy." But their story goes on, and it points out indicators of potential success that may serve other regions as well: Bernard and Young note that "this kind of resource management is home-grown and mindful of the need to sustainably use natural resources— rangeland and timber specifically"; "it welcomes partners, particularly folk who for generations have made their living from the land and waters"; "it strives to make decisions on sound scientific information and local knowledge of place"; and "it respects the web of living things and perceives that human well-being depends on the well-being of ecological processes."[15]

One way of looking at *The Ecology of Hope* is to see the whole book as an ecosystem that provides a habitat for healthy stories, a larger story made up of many smaller, interrelated stories, all of them stories of hope in which communities in desperate straits come together, driven by their very desperation, to work on profound and divisive issues. Bernard and Young conclude the story of the Mattole by writing, "Out of such work comes confidence that the valley may again become good salmon habitat and continue to be a healthy environment for people, too." Such stories make us think it can happen here—wherever you and I are—too.[16]

Those of us invested in trying to create a sustainable culture find a peculiar unity throughout creation. Richard Nelson explains that one of the underlying principles of the Koyukon worldview is that the natural world and the supernatural are one and the same.[17] In our contemporary American culture, we have become so accustomed to taking things apart that we have taught ourselves to see life in dichotomies and fragments. But the idea that all is one is reflected in the works of both Eastern and Western thinkers as we move back through time toward our common roots. Pre-Socratic philosophers and the earliest Zen priests appear to share a common view of the "earth breathing in, breathing out," a single organism that, like an animal or an ecosystem, is affected in every part when touched in any part. They seem to foreshadow the old Lakota song, "We are all related; we are all one"; they tell us we cannot harm or disrespect any part of the whole without ultimately harming and revealing our disrespect for ourselves. Stories are one of the most powerful ways we have to explore such an idea.

Sometimes that healing story is caught in a poem, as in these lines by Richard Hugo that could describe the Mattole process. In "Distances," he begins with the observation that "all things come close and harmless / first thing this morning, a new trick of light." The observation then becomes an image of hope.

> Let's learn that trick. If we can it will mean
> we live in this world, neighbor to goat,
> neighbor to trout, and we can take comfort
> in low birds that hang long enough for us
> to read markings and look up names
> we'll whisper to them from now on.[18]

One feels that this whisper is not just from fear that a louder voice would drive them away but that it is the soft whisper of respect, awe, or reverence.

These are the stories that will make us whole. In the telling and re-telling of them, we will grow into them, and our spirits as well as our spiritual lives will be enriched and enlarged thereby. Through language, through stories, we can write a geography for the world, not just about it. I have maintained throughout that we will not get far with changing the world till we make some changes in ourselves. Changing the stories we tell ourselves is part of that, and it takes both intellectual and spiritual discipline to create such stories.

Spirituality

The Power and Pragmatism of Language

It is about 490 B.C. Confucius has been traveling around China, teaching, hoping to find a ruler who will not only seek his advice about running the country but actually implement it. He has had little success. A practical man, a disciple—one suspects he had ties to political insiders—comes to Confucius, asking, "If the Lord of Wei wanted you to govern his country, what would you put first in importance?" The question has the feel of a test question, a hidden agenda left unspoken and lurking behind the interrogative: "Answer this right and I'll put in a good word for you." Confucius doesn't need time to think; he already knows his answer: "The rectification of names," he replies, "without a doubt." The practical man is astonished, as would be any presidential advisor in our own country and time. "That's crazy!" he replies. "What does rectification have to do with anything?" Confucius does not gladly suffer fools and does not coat many pills. For one whose whole life was given to political science, he is often impolitic in his speech. Truth came before schmooze. "You're such a dolt!" he says, then continues, "Listen. If names aren't rectified, speech doesn't follow from reality. If speech doesn't follow from reality, endeavors never come to fruition. If endeavors never come to fruition, then Ritual and music cannot flourish. If Ritual and music cannot flourish, punishments don't fit the crime. If punishments don't fit the crime, people can't put their hands and feet anywhere without fear of losing them."[1]

Without its right names, the world, as Confucius points out, is unreal, and none of the government's policies will be realistic, and none of

its endeavors will come to fruitful conclusions. In noting the importance of ritual and music, Confucius echoes indigenous peoples around
the globe, for those are two necessities for balance and harmony within
oneself and in the society. I take the final comment, "If punishments
don't fit the crime . . ." as a way of saying that people will have no confidence in government if the government does not call things by their
right names. One might commit a small crime, not even theft, and lose
a hand like a real thief caught with the goods. Without right language
we have government by caprice, which may be just the way some (many?
most? all?) people in power—princes of commerce and industry, members of Parliament and Congress, among others—want it. But Confucius apparently believed what the founders of our own government
often said: if the government does not have the support of the people, it
cannot stand. So the most critical arena for the implementation of successful, long-lasting government, in his view, lies in getting its language
straight. "Naming enables the noble-minded to speak, and speech enables the noble-minded to act," says Confucius. Part of his definition of
the noble minded is that they seek to square their acts with their words.
"Therefore the noble-minded are anything but careless in speech."[2]
How right he seems in our era of government euphemisms and deceit,
our lack of the noble minded in our leadership, their determined efforts
to use words to gain a chosen effect rather than to get them right.

Thus we begin our thinking about spirituality with the idea of a
right name, a clear word. My assumptions are that a healthy spirituality
is as deeply rooted in language as in faith and that speech is both cosmic and earth oriented as well as personal. I am also looking for a pattern here, beginning with the idea of clarity in the language, one of the
closest and most accessible of our human endeavors—and one of the
most complicated.

For those who are Christian, everything, at least since around A.D.
110, begins with the word. The word is logos, and it gets its most artful
and meaningful expression in the very first sentence of that Greekblessed writer of John's gospel: "In the beginning was the Word, and the
Word was with God and the Word was God" (John 1:1–3). The logos

since then, for those rooted in the Christian tradition, has been taken for the ultimate creative power, and for many it has been synonymous with God or Jesus.

But there is an older logos than John's, and his is but a descendent of that older logos. Half a millennium or more before John—about the time Confucius was talking about the importance of a clear word—Heraclitus, in the first fragment of his we have, says,

> The Logos is eternal
> but men have not heard it
> and men have heard it and not understood . . .
> Through the Logos all things are understood
> yet men do not understand
> as you shall see when you put acts and words
> to the test I am going to propose:
> One must talk about everything according to its nature,
> how it comes to be and how it grows.
> Men have talked about the world without paying
> attention to the world or to their own minds,
> as if they were asleep or absent-minded.

In fragment 32, Heraclitus goes on: "Thinking well is the greatest excellence: to act and speak what is true, perceiving things according to their nature."[3]

"Logos" is Greek for "word," the creative and active power behind the universe. Classics scholar Hazel Barnes once told me that "logos" for Heraclitus also meant "relationship"; it is our tie to everything else. Seyyed Hossein Nasr, an Islamic professor of comparative religion, takes that idea a step farther; in *Religion and the Order of Nature*, he tells us that "logos" also means "harmony." We are created not only for relationships but for harmony within ourselves, among ourselves, and between ourselves and all other creatures. Paul Woodruff, a classics scholar at the University of Texas, would agree with Barnes and Nasr that Heraclitus's notion is far greater than "word" as language or vocabulary. Over lunch in Austin, Paul told me that the nearest Latin translation would be "ratio," which has to do with reason, the rational, but also with ratios as in mathematics. The ratios that concern Heraclitus are those that are held in balance as the elements move up and down between heaven and earth. These ties among the word, the earth, heaven, relationships, and harmony fill in part of the complex picture of the logos in Heraclitus's view. Heraclitus also seems to say that it is the logos, the word, that establishes and maintains our relationship to the real world: "Man, who is an organic continuation of the Logos, thinks he can sever that continuity and exist apart from it."[4] He does not need to add, "How foolish!" If we break the word, we cut ourselves off from the natural; we break our ties with the world and one another, our bond with the only world that can sustain us. If the language of our public discourse and personal conversation does not give us the right names for things, then how do we rediscover and regain what is real?

Before the logos, Andre Padoux writes, there was in Sanskrit the word *vāc*, which represented another creative power, the "mother of the gods." It stood as "symbol for the Godhead" and revealed a "divine presence within the cosmos, as the force that creates, maintains, and upholds the universe." "Vāc," Padoux tells us, is translated as "word." "Vāc" was always an oral, aural word. An alternative translation is "speech," but this word "is an energy" that can be "tapped and used by anyone who is able to penetrate its secret nature and mysteries," explains Padoux. One of those mysteries is that, though "vāc" can be translated as "word" or "speech," as a creative force it comes before language and cannot be translated as language. In ancient India this word indicated a "constant ambivalence" about the nature of the human and the cosmic. The "knowledge of the supreme reality, the highest understanding, was founded on knowledge of anthropo/cosmic correlations," writes Padoux. So there was "no distinction between the human and the cosmic, the vital, the psychic, or the spiritual."[5]

Nasr also points out that in ancient Egyptian theology, the divinity Ptah contains the creation in his heart and creates beings through his tongue, that is, by his word. The great cosmic realities, called the Ennead—including Atum, the primordial Adam—are created by Ptah's simply pronouncing their names.[6]

The power of the word in all these traditions is related to the power to name things. In fact, the name of a thing and the thing itself seem to be the same. Thus John's gospel tells us that the word is not only with God but is God. Extending the dimensions further, the Upanishads hold that the name is conjoined not only with the thing it names, but with any action that is undertaken in that name. Thus the *Brihad-Aranyaka Upanishad* reads, "The Universe is a trinity, and this is made of name, form, and action. Those three are one." This whole notion takes a very human creative act, the act of speaking, and gives it primacy, says Padoux, then "chooses to reverse the order of things" so that the word comes first and our human speech is simply the continuation of a divine, cosmic process that continues forever. These comments echo the Yup'ik: "Through the naming process, the essence of being human is passed on from one generation to the next," Ann Fienup-Riordan explains in *Eskimo Essays*.[7]

In *The Great Digest*, Confucius seems to say that clarity in the word—not the purity that guardians of grammar seek but the precise word—is the root of integrated persons, strong families, and good government. "If the root be in confusion," he says, "then nothing will be well governed." The precise word is the result of the widest and best learning possible, and it becomes the source of all self-discipline and self-cultivation. Ezra Pound translates the Chinese characters for the accurate word as "the sun's lance coming to rest on the precise spot verbally," as if the clear word will bring anything to light, even the very heart of our most inward self.[8]

The *Tao Te Ching* opens by insisting, "The Tao that can be told is not the eternal Tao. The name that can be named is not the eternal name. The nameless is the beginning of heaven and earth." Yes, there is a certain contempt for language built into those lines, and it conflicts with Confucius's concern and appreciation for language, even though words cannot do everything we'd like and may be as often imprecise as accurate. Yet even in that Taoist insistence on a way that transcends language, there is room for the clear, creative word, for the next line is, "The named is the mother of the ten thousand things."[9] The ten thousand things are of a lower, noneternal order—the immediate, contingent, immanent, earthly stuff in all its manifestations that surrounds, perplexes, pleasures, elevates, expands, sustains, and tempts us, all those created things that compose our world—nevertheless they seem to come from a word. How do we get from the nameless to the named without a word slipped in there someplace, somehow, between the eternal and the immediate? The eternal is uncreated, I take it; the ten thousand things are created, but by what or by whom, how? Perhaps by what names them? I admit there is room for more than one thing between the eternal and the things around us—a divine agent, perhaps, who gives names, and the names themselves. But if the name is the agent, then the naming word has a role here, if not as a being then as an action or a force that gives form to the ten thousand things, quite like the logos or the Upanishads—or the Chinese *qi*, a word for the creative principle that holds the world together, regenerates us anew from the losses we all face, and offers us the benevolence of fecundity. If the ten thousand things come to us by their naming, and if they are as complex, linked, united, related, all one, as our own contemporary observations lead us to believe, then the naming in the *Tao Te Ching* is a powerful force akin to the power of relationship and harmony that Heraclitus says binds us together in the right ratios. And if the creative agent that gives us the names is language or the word, does it matter if the word is logos or vāc or something else?

Augustine, the obsessively self-cultivating bishop of the early church, emphasized the importance of the oral word for teachers when he said, "And so I learned, not from those who taught, but from those who talked with me." James Hillman, a psychiatrist, tells us in *Healing Fictions* that the avenue to wholeness and integrity lies in keeping our stories straight, standing by our word. Pushing the notion even further, Victor Frankl, a Holocaust survivor and creator of a whole psychiatric system, named his approach to creating a healthy psyche "logotherapy."[10]

We have not followed John or Confucius or Heraclitus, or listened to the goddess Vāc, and have thus made Hillman and Frankl and all our psychiatrists, shamans, and *curanderas* necessary. Bill McKibben points out in *The End of Nature* how serious breaking our bonds with the earth may be.[11] Jinx Everett, a Colorado schoolteacher, told me in a teachers' workshop that he thought the crooked language of government and the media might be an even more dangerous threat to our society than the current ecological threats to our environment.

The Pragmatism of the Word

All that may sound pretty abstract, even ethereal, and many of us are impatient with abstractions. We turn away from them because they are too vague, not pragmatic enough, or because they do not appear to offer an immediate return. We are like the practical man who came to Confucius; we don't believe we can take abstractions back to the prince or the office or the classroom and *do* anything with them. But I want to reclaim the value inherent in certain abstractions, including the idea of the word. Often we want to "get on with things," "get up to speed," get away from the conceptual and lose ourselves in action. How we speak about things, then, is rendered unimportant. "That's just semantics," we are told, and semantics are seen as an obstacle to action. It is possible to pick up speed much earlier if we bypass the language of conception. What that most often means, however, is that the wreck waiting for us down the road will be a fatal one instead of a fender bender.

Mathematics is an abstraction, yet we do not ignore or dismiss it, and when things finally add up, we see its value. We see government as an abstraction until we attend our first precinct meeting or party caucus, where it can quickly become very personal, demanding our thought and our time. We see the law as an abstraction until we begin to apply it, or a uniformed officer begins to apply it to us. We see the idea of the word as an abstraction—until we break it, or put it in the service of unworthy ends. I am utterly with Confucius in this. I do not believe for an instant that the idea of the word is an abstract notion, for clear language is critical to the development of trust, the maintenance of social cohesion within as well as between societies, and between citizens and the governments that represent them. But I'd have to confess that my own approach to this is based in my experience and is therefore per-

sonal, perhaps idiosyncratic, and can't be blamed on Confucius or anyone else. Neither does it come from the multitude of deceits promulgated
by our government with increasing frequency. Where's Confucius when
we need him?

America has always been known as deceitful. In 1897, Sitting Bull
informed a reporter for the *New York Herald*, "I have told my people
that the Americans are great liars." We seem to think, because we
Americans have such restricted access to the world's news (or did have
until very recently), that the rest of the world does not know that we
were lying about Nicaragua, Panama, Guatemala, Grenada, Chile, El
Salvador . . . Afghanistan, Iraq. But they did know; they knew it when it
was happening. The unreality of our word has become so egregious recently that no one, even those nations whose aims and means are as
duplicitous as our own, trusts us. No wonder. Our deceits started long
ago, and they will not end soon, for they have never been limited to a
single political party. But all that is global, so let me spell out a couple of
personal experiences that lead me to believe that the idea of the word is
not abstract but pragmatic and, despite the lure of realpolitik, is essential to a sustainable government that is the expression of a sustainable
culture.

In February or March 1986, I was in Nicaragua, standing on a
porch in the dusk in La Paz del Tuma, an *asentamiento* in the mountains beyond Jinotega. Oscar, a Nicaraguan from Managua, and I were
simply staring across the big valley, watching the mountains across the
way looming black and beautiful against a rapidly darkening sky. Whatever thoughts were absorbing Oscar, I was practically mindless, soaking up an evening of sunset clouds and shadows and the mystery of
being in this darkening, alien place. A red arc flashed against the for-
ested background, and then others, all so far away there was no sound.
"Tracers?" I asked, and Oscar replied, "Sí, tracereros." They were far off
to the east, not a threat to us and probably not even a firefight, just Nica
soldiers or contras sounding off, letting the adversary know they were
around. If it had been serious, the tracers would have been low, flatter,
obscured by the jungle. Then the noise of guns and the red streaks of

other tracers came, but closer and from the north. From the west, louder and closer still, came the thump of a mortar, the rattle of automatic weapons, no tracers visible. None of these were yet a danger to us, but the night was suddenly alive with virulent sound, the popping, booming language of hate, filled, for some folks not far away, surely with terror, and perhaps with pain or death. Idyllic reveries no longer seemed either possible or appropriate.

In all, as many as seventy thousand Nicaraguans died—out of a population the size of Wisconsin, over half of whom were under sixteen years old—in a terrorist war against peasants whose government, for the first time in their lives, allowed them to send their children to school, provided access to health aides, and taught both children and adults to read. That contra war included the deliberate rape of daughters in front of parents held at bay with automatic weapons supplied by the United States. Some of those fathers were flayed alive and left in the jungle to rot. The contras were funded and trained by the United States, whose president was spread across the U.S. media holding a T-shirt that read "I'm a Contra too!" and who insisted the contras were "the moral equivalent of the Founding Fathers."[12]

The only place I know I was lied to in Nicaragua was in the U.S. embassy—I knew the facts were different from what the embassy people told me and that they must have known they were different. It was also the only place in that strife-ridden country, including military sites, where my camera and tape recorder were confiscated before I entered. Secrecy not openness, control not freedom, deceit not truth was the order of the day at the embassy. The U.S. personnel there must have assumed that our Witness for Peace group was extremely ignorant or perhaps that we could not read or hear, for what they told us were old stories, long since disproved. They repeated the tale of a "massive flow of arms from Nicaragua to El Salvador," which we had heard from Reagan, while the CIA was admitting that there were no such shipments, that the arms from Nicaragua to El Salvador were so small they could have come from someone's garage. When we asked about events in Escombray, a remote *asentamiento* we had visited, which was threatened with a contra attack the day we left, the embassy spokesperson said that he had never heard of the place and that they were confined to the embassy—they could not travel in the countryside because Congress would not give them four-wheel-drive vehicles. But in that dry season we had driven all over the most difficult roads, one so steep we had to get out because our van, a dilapidated old two-wheel-drive Toyota, did not have enough power to pull it with us inside. The embassy spokesperson also indicated that travel in the countryside was too dangerous, that Sandinista soldiers guarded all the bridges and would not let civilian vehicles pass. Yet we, and people in many other vehicles, had cruised across many such bridges, without slowing down and without receiving even a glance from the guards.

The real threat in the countryside was not Sandinista soldiers but American-made, American-supplied antipersonnel mines buried in the roadside, the same strategy now used by "terrorists" in Iraq to kill our soldiers in their Humvees. In Nicaragua the mines were used by the American-trained and American-supplied contras not to kill Sandinista soldiers but to harm civilians. The purpose was not to engage in combat but to cripple the economy by terrorizing the countryside so residents had to neglect coffee picking and flee to the cities for protection. In the city, the real threat to us was presented by the embassy personnel, who were not out to inform their fellow U.S. citizens but to hoodwink us. When we left the compound, the large brass eagle, symbol of America, screwed on the iron gate suddenly looked threatening and vicious, as I had never seen it before. It is hard not to think that the United States wants it that way.

There was an active communist party in Nicaragua at the time. It was opposed to the Sandinistas and had little public support. The differences between the Communists and the Sandinistas were clear, and they were made clearer by the Communist Party officials we talked with. But the differences were deliberately obscured and manipulated by both our press and our government. Nicaragua's appeals to the Soviet Union were forced by U.S. policies and embargos. The United States refused to help; even medicines were embargoed. The Sandinistas could let disease ravage their people, let their people starve and the fairly elected government collapse, or turn elsewhere for help. They turned to the other superpower of the day, and the cry of "Communists!" rose from our government and our press immediately.

In Managua I watched the mainline reporters, mostly hanging around the one international hotel, taking government releases from the U.S. embassy and printing them without investigation, never venturing into the countryside for a firsthand look. They treated the government handouts as if they were news instead of disinformation, truth instead of fabrication. Those who did get into the countryside and tried to write a realistic account of events could not get their stories into print or, if they did, were soon out of work.

I came back from Nicaragua a little crazy, depressed for months, not from any threats to my personal safety but because I felt totally betrayed by my government and its media, knowing from my own experience I had been lied to by both. I thought for a time that if I read one more lead paragraph in the *New York Times*, the *Washington Post*, or the *Los Angeles Times* that began "The communist Sandinista government of Nicaragua," I would literally be sick at my stomach. I've hardly read a mainline newspaper from then till today. If they were so deceitful about this war, why would I trust them about other things? Now I find what I need to know from other sources, more trustworthy, mostly from firsthand accounts of friends who continue to go to Central America regularly. The anecdotal observations of friends are generally more insightful, better informed (they speak the language and have been studying and visiting the region for years), and more accurate than the "objective" reports of professional journalists and government employees. I also turn to small, mostly "radical" journals, some of whose reporters used to work for the Associated Press but could not get true stories printed because they didn't square with either the publisher's or the administration's view or policy or propaganda. That's one reason why the word, a glowing arc across cultures and eras and falling into our own lives, seems to me not abstract but very personal and pragmatic.

In December 1986, I was in Iraq. Basra, in the south, was taking thousands of rounds of Iranian shells every day. After spending a day at the front in the north, I returned to my hotel in Baghdad and was changing my clothes when I heard a swoosh and blam! I stepped out on the balcony to see what had happened. Across the Tigris, perhaps 250 yards away, a black column of smoke was lifting where an Iranian missile had blasted into the base of an apartment house just at the end of the bridge I crossed every day to visit the souk. People were yelling loudly enough to be heard, even at my distance, and the sirens on emergency vehicles were already shrilling. Two days later a missile landed within a block of the hotel. I was up north in Mosul and Ninevah when it struck, but when I got back, a friend who'd stayed behind told me how close it was. A few days later, another fell, farther away.

The sporadic, targetless randomness of those missiles simply being lobbed into the city to hit whatever lay before them was more chilling than if they had been more deliberate or more frequent. The Iranians—and we—apparently just wanted to kill a few folks, blow something up, it didn't matter much who or what it was. This was war, after all. And I say "we" because, like the land mines in Nicaragua, those missiles falling into Baghdad apartments were American made and American supplied. By coincidence, that was the week the Iran-contra scandal unfolded in Middle East newspapers and then exploded in headlines around the world. Till that moment, most of the world had thought, based on American public relations rhetoric, that the United States had been leading a crusade to impose a worldwide embargo on arms shipments to Iran, making much of the fact that no one should support the war by helping Iran with parts or ammunition or weapons. Yet all the while we were doing it clandestinely ourselves, and Colonel Oliver North, contemptuous of the Constitution, was lying to Congress, America, and the world, breaking his word at every turn. I had learned earlier that week in Baghdad, from an article published for British readers by a BBC correspondent, that the upper levels of the Iraqi government knew that the American embargo was a lie, but most folks in Iraq and in the United States didn't. I will never forget the dismay, the sense of betrayal, and finally the anger felt by Iraqis when they learned that we had been sending Iran war materials all along. Both sides had ample cause to despise us.

As a participant in a war between Persian and Arab cultures, the United States claimed as part of its public justification that it was trying to maintain the balance of power in the Middle East. We could see it as balance, I suppose, because while we were sending parts and arms to Iran, we were also sending satellite information about Iranian troop deployment to the Iraqis, letting Iraqis think we were supporting their efforts. It turned out, finally, that the missiles didn't work all that well and that we had doctored the satellite information before it was forwarded to Iraq. Thirty-six months later, we were calling Saddam an impossible liar.

For the second time in a year, I was far from home, betrayed by my own country, my life threatened this time, not by violence-prone "fundamentalist foreigners" conjured up by our media but by my own government, the hypocrisy of its rhetoric, and the double or triple standards implicit in its policies, obvious to everyone in the world except Americans. I was suddenly confronted again by the broken word, the breaking of a tradition that stretches from beyond Heraclitus and Confucius through John's gospel to me, and finding my life threatened by it.

Fifteen years later, on September 11, 2001, similar horrors happened three times, in a matter of minutes, in the United States. But because our media had not given us much hint of our actions around the world, we saw our citizens as innocent and were baffled by banners in other countries proclaiming that Americans should think about why so many in the world hate us.

Don't tell me that if I say such things I don't love America; I do. And don't tell me it's our country right or wrong. That only multiplies the world's wrongs, and it has quite enough, thank you. America's landscapes, its people, and its ideals are as beautiful as any in the world. But, because of the words it has broken so often, I have learned to love America as one loves a spouse who is not always faithful. There is pain in the loving. And when we discover secret liaisons, as those with Iran in the 1980s, or with a shadow government in our own country accountable to no one, the pain is double, for we know that in America's broken word it betrays not only us but itself as well. I'm not asking you to buy my politics or to agree with my analysis; I'm just telling you where I'm coming from and why I think abstractions like the word need ample thought and attention. They have a power that reaches far beyond abstraction, a power that reaches deeply into our daily lives and, when broken, may take them.

In our time the word "terrorism" operates pretty much as the word "massacre" did in the nineteenth century. When whites killed Indians, it was a skirmish, a battle, a military operation; when Indians killed whites, it was a massacre. Now, when our enemies bomb us or hijack a plane, it is terrorism. When we inflict terror on rural Nicaraguans, it is a "low-level war of attrition." When Israel invades Gaza, murders innocent victims, terrorizes and even deliberately shoots children, the Western world remains silent because Israel says it is fighting terrorism and is, after all, our ally. When a Gaza suicide bomber kills shoppers and children in Israel, we do not understand how a terrorist might do such a thing. When it comes to rogue violence, neither side can claim virtue. We might have had a chance at it if our educators had read their *Meno*.

So Iraq is the second reason that for me the word, broken, is as pragmatic as a piece of shrapnel or the red arc of a tracer. Together, the two reasons become one vital indicator that America is not sustainable and cannot last one-tenth as long as some Eskimo villages already have. Their indigenous longevity is not attributable to superior arms or technology— though their technology includes some of the sleekest marine engineering ever devised by humans—but to keeping their stories straight. Confucius was right: for our purposes in thinking about how to create a sustainable culture, even a sustainable United States, safe and productive over time, the issue of language is paramountly pragmatic. Using language to get at the truth is more enlightened than using language for its persuasive effect, as we have learned to do so well in our public relations factories, especially if that effect supports a lie. Our heads spin trying to understand the world in part because the political rhetoric in every nation spins. These personal experiences are, in part, why I have come to believe that a healthy contemporary spirituality lies in our language.

In a House Made of Dawn

In *House Made of Dawn*, N. Scott Momaday's wonderfully evocative novel about Pueblo people in New Mexico just after World War II, Abel, the protagonist, has just returned from the war, disoriented by his experience and unable to take part in the traditional rituals. He has become angry and hard. Though home physically, he is lost and wandering in his soul, unable to return to the real life of his people. At the heart of his problem is the loss of words. Momaday describes the dilemma as follows:

> His return to the town had been a failure, for all his looking forward. He had tried in the days that followed to speak to his grandfather, but could not say the things he wanted; he had tried to pray, to sing, to enter into the old rhythm of the tongue, but was no longer attuned to it. And yet it was there still, like memory, in the reach of his hearing, as if Francisco or his mother or Vidal had spoken out of the past and the words had taken hold of the moment and made it eternal. Had he been able to say it, anything of his own language—even the commonplace formula of greeting, "Where are you going"—which had no being beyond sound, no visible substance, would once again have shown him whole to himself; but he was dumb. Not dumb—silence was the older and part of custom still—but *inarticulate*.

After a long walk in the canyon, "he began almost to be at peace. . . . He was alone, and he wanted to make a song out of the colored canyon . . . but he had not got the right words together."[13]

Later, after killing a man who was trying to kill him, after being in jail and being beaten severely himself, Abel is in Los Angeles, desperate and alienated from his prewar reservation world, still without his language and the ceremonies and rituals that might have brought healing to his wounded spirit. An observer on the street, seeing Abel drunk, might assume that his problem is alcohol, but it is not drink that has disabled Abel; it is his inability to rectify the names, his being out of touch with the right words. He has lost his tongue. In the heart of the great city, he goes one night to a church meeting led by Tosamah, a street preacher who calls himself the Priest of the Sun. Right in the center of this book about Indians there is this sermon, preached by a Navajo who is a hip, swinging street person, like many in his congregation. He begins,

> Good evening blood brothers and sisters, and welcome, welcome. . . .
> "In the beginning was the Word," I have taken as my text this evening the almighty Word itself . . . and the riddle of the Word, "In the

beginning was the Word . . ." Now what do you suppose John *meant* by that? That cat was a preacher, and well, you know how it is with preachers; he had something big on his mind. Oh my, it was big; it was the *Truth*, and it was heavy, and old John hurried to set it down. And in his hurry he said too much. "In the beginning was the Word, the Word was with God." It was the truth all right, but it was more than the Truth. The Truth was overgrown with fat, and the fat was God. The fat was *John's* God, and God stood between John and the Truth. . . . That cat had a whole lot at stake. He couldn't let the truth alone. He couldn't see that he had come to the end of the Truth, and he went on. He tried to make it bigger and better than it was, but instead he only demeaned and encumbered it. . . . He imposed his idea of God on the everlasting Truth. "In the beginning was the Word . . ." And that is all there was, and it was enough.

The Priest of the Sun has had enough experience with white Americans to know their non-Indian culture very well and to understand some of the ways they deviate from their best traditions. His words point up the primary peril in our own efforts to create a sustainable culture. It is exactly the one named by Confucius 2,500 years ago: for our American culture to be sustainable, we have to rectify the names. Red Cloud couldn't have said it better.

Now brothers and sisters, old John was a white man, and the white man has his ways. Oh, gracious me, he has his ways. He talks about the Word. He talks through it and around it. He builds upon it with syllables, with prefixes and suffixes, and hyphens and accents. He adds and divides and multiplies the Word. And in all of this he subtracts the Truth. . . .

He has diluted and multiplied the Word, and words have begun to close in upon him. He is sated and insensitive; his regard for language—for the Word itself—as an instrument of creation has diminished nearly to the point of no return. It may be that he will perish by the word.

If that is to be our fate, we ourselves will be the cause of our own undoing; we will have succumbed to our penchant for putting our words into stories that are not true, or whose unhealthy purpose is short-term gain rather than long-range sustainability. The Priest of the Sun eventually concludes his sermon this way:

Do you see? There, far off in the darkness, something happened. Do you see? Far, far away in the nothingness something happened. There was a voice, a sound, a word—and everything began . . . Say this: "In the beginning was the word . . ." There was nothing. There was *nothing*! Darkness. There was darkness, and there was no end to it. You look up sometimes in the night and there are stars; you can see all the way to the stars. And you begin to know the universe, how awful and great it is. . . . The darkness flows among the stars, and beyond them forever. In the beginning that is how it was, but there were no stars. There was only the dark infinity in which nothing was. And something happened. At the distance of a star something happened, and everything began. The Word did not come into being, but *it was*. It did not break upon the silence, but it was *older than the silence and the silence was made of it*. . . .

And from that day the Word has belonged to us, who have heard it for what it is, who have lived in fear and awe of it.

Tosamah also tells us in his sermon that, for the Navajo, the name of the original state of the universe is "One Word." In another context, Momaday writes, "A word has power in and of itself. It comes from nothing into sound and meaning; it gives origin to all things."[14]

Abel is finally restored, reunited within himself and with his people's ceremonial life, whole again. He has achieved two amazing transformations: he has relearned compassion and found restoration to the tribe. He has arrived home from the war at last. Both those transformations depended on his finding the right words and on another ancient formula for regeneration and transformation, a formula found in the ancient Chinese and in every American Indian tribe we know of: participation in appropriate rituals. He recovers his people's tongue and is free to join the lasting ceremonies of the tribe. In the participation he is made whole. His grandfather, whom Abel has saddened, embarrassed, and betrayed, dies, and Abel is compelled to return to him. "He knew what had to be done" for his grandfather, and he does it. He prepares his grandfather's body according to the tradition and sees to his proper burial. Then he goes out to the stick race, another tradition, and he begins to run, taking part in the great ceremonial. He is not running to win anything; "he was running and there was no reason to run but the running itself and the land

and the dawn appearing." He runs through exhaustion and pain, and he regains his old vision and the old ceremonial words, for the roots of compassion and restoration often lie in beauty and pain. When Abel begins to run in the prologue to *House Made of Dawn*, he is still not entirely back in his old world. The prologue ends, "Against the winter sky and the long, light landscape of the valley at dawn, he seemed almost to be standing still, very little and alone." But now, at the book's conclusion, Abel is restored to ceremony and to the land. His clouded vision of himself pales as he runs. He can see again, and he finds his voice; the right words come to him as song. "Pure exhaustion laid hold of his mind, and he could see at last without having to think. He could see the canyon and the mountains and the sky. He could see the rain and the river and the fields beyond. He could see the dark hills at dawn. He was running, and under his breath he began to sing. There was no sound, and he had no voice; he had only the words of a song. And he went running on the rise of the song. *House made of pollen, house made of dawn. Questaba.*[15] With the novel's last words, he has run through the pain, not the pain of running but of living, and he has found the right words again, and nothing is said now about his seeming "almost to be standing still," and nothing is said about his seeming "very little and alone." He has become not larger but the right size, filled again with tradition, with the right words, with ceremony and song, in his proper place in the universe. Such is the power of the word to balance and harmonize heart and mind, to unite and to heal.

Speaking from deep within his culture's own distinct tradition, Gregory Cajete, a Tewa Indian educator, holds that one of the critical concepts commonly held by American Indians of many tribes is that of breath. It is breath that holds the world together, Cajete writes in *Look to the Mountain*, and breath that provides the link between all things. That is less than half a step away from two notions included in Heraclitus's logos, for inherent in the breath of the American Indian is not only the idea of relationship but also the idea of harmony. For Cajete, words, then, are "the sacred expression of the breath," and one hears the old Egyptians and the traditions of other cultures echoing through his words. "The subtle truth," says Lao Tzu, "can be pointed at with words, but it can't be contained by them." Then, sounding much like Cajete in his description of the American Indian tradition, he adds, "The breath of the Tao speaks, and those who are in harmony with it hear quite clearly."[16]

Simon Ortiz, an Acoma poet and storyteller, says that "words are so fundamental in themselves, that they cannot be broken down." But, Ortiz warns us, getting our words right is not only a matter of grammar: "One has to recognize that language is more than just a group of words and more than just the technical relationship between sounds and words. Language is more than just a functional mechanism. It is a spiritual energy that is available to all. It includes all of us and is not exclusively the power of human beings—we are part of that power as human beings. . . . We forget that language beyond the mechanics of it is a spiritual force."[17]

Getting our words straight is part of living in a sensible and sustainable universe, and the stories that are conveyed by those words are part of creating sacred relationships with all the others in the world around us. If we distort language, use it as a mere mechanism to gain power or wealth, our spiritual life becomes as twisted as our words. When things go sour and the world turns painful, the traditional healing ceremonies, the songs and chants, are aimed at restoring balance and harmony between oneself and the world—getting our stories straight. It becomes terribly important that the words of the songs and the chants be the right words. If they are the right words, they take on the healing power of the sacred forces of the universe, able to restore the harmony and the spiritual force Ortiz and Heraclitus and Confucius and John and Frankl and Hillman and Cajete all see as innate in the word. Consider the psychic blow that the loss of their language is to native speakers; then consider the loss of meaningful language all of us have suffered given the onslaughts made on it by journalism, commerce, government, and, yes, academe—a loss that has us all disoriented and untrusting.

Every scholar will hasten to say that Heraclitus's logos is not the same as vāc; that the Egyptian notion of breath is different from the breath of American Indians and that both those are different from the *ruah*, the "breath of God" as that phrase is used in the Old Testament, and from the *pneuma* of the Greeks. I find those differences important, and I do not mean to ignore them, but what impresses me even more is that people from all those differing cultures and centuries and disciplines—and I have not even begun to exhaust them in this small survey—elected similar metaphors for the ultimate mystery that surrounds us. They all recognize that something holds the world, with all its disparate parts, together. The images chosen by so many circle around word, language, breath, or speech, none of which is possible without the others, and together they form a kind of sacred music and mucilage, a gravity in and of the word, that holds everything together.

The message of all these folks seems to be that the clear word is the basis of balance and harmony in all things. Though this is not made explicit, in later works attributed to Confucius, clarity in the language, the precise word, may be what he calls the "pivot," the "axis in the center," which is also "the great root of the universe." The great person, in the view of Confucius, is the one whose word is always true, straight, unwavering, a plumb line akin to that of the Old Testament prophet Amos, against which both actions and words can be measured. We will never get our politics, our national psyche, or our relationships with one another and the earth harmonious until we establish that kind of clarity in the language. Without that clarity we will never approach becoming a sustainable culture.

The Priest of the Sun was right: we have not followed John; even John did not follow John. "The truth is fairest naked, and the simpler its expression the profounder," Schopenhauer says.[18] Nevertheless, the truth is there, and I am happy to take the word as the great metaphor for the mystery at the heart of all being, the focus of our spiritual quest, to believe that out of the word came everything we know. Assume for a moment that the word as metaphor is its reality; assume for a moment that Heraclitus is right and the Upanishads are right and the Priest of the Sun is right. Assume that the new story of the universe now being told by Brian Swimme, Chet Raymo, Martin Rees, and other scientists is right, that out of a climactic event, the big bang, the fundamental work was set in motion that resulted in the finely balanced relationships that created the world we know. Assume that the big bang was an expression—a word without a native tongue—of the creative power and authority of the cosmos. How can this all be? How can the one be true and the others be true also? Can we recognize an appropriate spiritual life for our time in those images? I am looking for a spiritual life that will lead us in our work of self-cultivation—the kind of cultivation that makes us full enough persons that we have something to give the world—and will lead us to recognize our real subsistence and the means to create a sustainable culture. I hold that one discipline that nourishes the spiritual life lies in our language, in clarifying our words, rectifying the names, and that the use of metaphor, rather than obscuring meaning, may well strengthen not only our language and stories but also our spiritual life.

Clarity and directness in the language are not simply intellectual efforts but spiritual discipline. Often the truth is dangerous for one reason or another: it may hurt others or reveal our own shameful or embarrassing actions. We know also that there are times when silence may be as damaging to our souls as speech. Language then is an immediate ethical issue, and the choices we make about our words either nourish the soul or diminish it. Assume, too, that Ortiz is right: beyond the mechanics, language is a spiritual energy available to all. Assume, again, despite all the evidence that clarity in our words is not an operating principle for our culture, that Confucius is right, and the first task of government is to call things by their right names. And assume that Epictetus is right that perfect speech marks a moral life. That's a lot of assumptions, perhaps too many for our skeptical culture. Yet the evidence is abundant that the clear word, for many cultures, including our own, represents a profound spiritual searching of self and concern for others. And it becomes the germ, the seed, the pivot point, the source of the relationship between Nature and our own human spirituality.

Rectifying the Names

IF YOU CAN SUSPEND disbelief, as required by every drama, then believe this for a moment: Out of the word came everything we know. Nature. The world. The earth. All creatures, plants, peoples. Four-leggeds and two-leggeds. Wings, legs, heads, thoraxes, abdomens. Thoughts and visions and dreams. We are considering notions regarding Nature and spirituality, but when I hear many people nowadays talk about spirituality, I am puzzled. It seems as if the current definition of being spiritual simply means that I think about my psychological health, that I take refuge in Nature, that "I understand Native American spirituality and where Native Americans are coming from because I once sat in a weekend retreat with a genuine Native American shaman," that I get touched by the sight of birds or the sun setting over the ocean or mountains or cornfields.

This is nothing new, although it seems more pervasive and simplistic perhaps than it did in earlier times. Evelyn Underhill begins her 1936 book *The Spiritual Life* by noting, "The spiritual life is a dangerously ambiguous term; indeed, it would be interesting to know what meaning any one reader at the present minute is giving to these three words." She continues, "Many, I am afraid, would really be found to mean 'the life of my own inside.'" Others would think of it as "something very holy, difficult and peculiar." She concludes wryly that this latter view sees spirituality as "a sort of honours course in personal religion—to which they did not intend to aspire." Both those views in Underhill's opinion are too individualistic and self centered. We need to get away "from those petty notions," she holds, and focus on "that great spiritual landscape which is so much too great for our limited minds to grasp."[1]

I am feeling my way along here, on thin ice cracking all around me, testing an area about which I feel inadequate . . . and yet the current spiritual scene seems much too facile, too rootless. Though it appears to take Nature into account, it is not genuinely grounded and has no commitment to hard thought beyond the current conventional wisdom, which is generally more current and conventional than wise. It is certainly possible to separate spirituality from conventional religions, or from religious institutions, but it is not possible to separate spirituality from the religious. Spirituality transcends all religions; we all know persons who are profoundly spiritual but belong to no particular church or faith. This does not mean they are irreligious. Neither is it possible to separate the spiritual from Nature and the natural. Because I am on thin ice and seeking (what—safety? a way out? a firm path?), I must look at as many of the available resources as possible, hunting down usable bits that will make a whole mosaic that is sensible, the fragments finally integrated into a larger picture.

Perhaps the problem is simply that spirituality is a complex idea or set of ideas. Robert Cummings Neville, in *Boston Confucianism*, agrees that "spirituality need not be closely associated with religions," but he, too, sees a tie between spirituality and the religious. He holds that there are three interrelated aspects to the religious enterprise: "ritual, cognition, and spiritual practices, all shaped by religious symbols engaging what is taken to be ultimate." For Neville, ritual "includes not only explicit liturgies" but also "repetitive behaviors" that people think are important reflections of our relationship to the world. The cognitive part includes "myth, cosmology and philosophy," which folks take to be basic categories of reality. Perhaps most important for our discussion, "Spiritual practices encompass the range of behaviors, corporate and personal, aimed at communal or personal transformation so as to better relate to what is taken to be ultimate."[2]

One reason we turn to other traditions, such as American Indian or

Buddhist traditions, is that we don't have many spiritual practices that work for us anymore. Our churches and our culture too often offer indoctrination: a closing down of possibilities, rather than of an opening of ourselves to possibility. But most important, we also seem to have lost the self-discipline to work on self and communal transformation. Indeed we may even believe that we no longer need transformation. If we do, we expect to find an easier way in other traditions. Yet self-cultivation is the great spiritual exercise in every great tradition, and in every tradition, spiritual practice requires the utmost in effort. In each of the great traditions, the purpose of self-cultivation is never simply to realize the self. It is to put ourselves in a better position to serve our families, our friends, our society, and the natural environment that surrounds us—and to put us in a better relationship to whatever it is that we see as the ultimate reality (whether transcendent or immanent in the cosmos). It is that improved relationship that finally makes us feel that we have found our place and are at home in the world.

Martin Buber writes, "Religiosity is man's urge to establish a living communion with the unconditioned; it is man's will to realize the unconditioned through his deed, and to establish it in his world. Genuine religiosity, therefore, has nothing in common with the fancies of romantic hearts, or the self-pleasure of aestheticizing souls or the clever mental exercises of a practiced intellectuality." Rather he finds it in our daily life and work: "Genuine religiosity," Buber says, "is doing." It lies in sculpting the unconditioned out of the matter of this world. For him, "the countenance of God reposes, invisible, in an earthen block." It is our human task to carve it out, let it appear from our work on the earth. "To be engaged in this work means to be religious—nothing else." He continues,

> Men's life, open to our influence as is no other thing in this world, is the task apportioned to us in its most inward immediacy. Here, as nowhere else, there is given to us a formless mass, to be in-formed by us with the divine. The community of men is as yet only a projected opus that is waiting for us; a chaos we must put in order; a Diaspora we must gather in; a conflict to which we must bring reconciliation. But

this we can accomplish only if, in the natural context of a life shared with others, everyone of us, each in his own place, will perform the just, the unifying, the in-forming deed: for God does not want to be believed in, to be debated and defended by us, but simply to be realized through us.[3]

There is a common, underlying element in Buber's contention, in the Upanishads, in the sermon by the Priest of the Sun, and in Heraclitus's gleaming fragments: the world is one. For a time we argued over whether everything in the cosmos is continuous or discontinuous. After Newton we insisted in our mathematics that everything is continuous—until physics showed us what our least sophisticated personal observations always led us to believe: that there are gaps; things do come in units. Whether continuous or with gaps, whole or fragmented, nothing comes without relationships, without connections. When those relationships are healthy—that is, when the things around us are healthy—we tend to be healthy. If the things around us tend to be weak or spindly or unhealthy, we have little chance to become otherwise. There is an astonishing unity that we share with everything else in the cosmos, animate and inanimate. That unity is established by our relationships, our ties, to everything around us. The ties of all those things we relate to extend their own connections to everything around them, an expanding circle of connections. Despite our individuality there is a sense in which we are all a tangle of ganglia, reaching out, ultimately touching all, everything related, everything one in our unified, interrelated (the word itself implies units connected) cosmos.

This entire, mysterious, ever emerging, ever puzzling cosmos is what I think Buber would call "the unconditioned." It is the primary material with which we have to work, and it is what undergirds, lies immanent within, our souls and is of the substance of God himself, herself, itself. In many traditional views, including the Western view, that unity is created and sustained by, indeed is, the word. Right here, embedded in this great mystery and metaphor, lie the clues we seek to the relationship between Nature and the spiritual life.

In the earliest Western visions of the universe, our elders sound like

Zen priests or some truly archetypal American Indian shaman, seeing the world as a single, living organism. Pre-Socratics like Thales, Heraclitus, Empedocles, and Diogenes are very clear about this. "Everything breathes in, breathes out. . . . So all creatures have a share of scent and breath. . . . All things have intelligence and a share of thought," writes Empedocles. In the beginning, he continues, "all creatures were tame and gentle to men, all animals and birds and loving minds glowed." Diogenes proclaims, "Everything is of one substance," but "we have complicated every simple gift of the gods." How much he sounds like the Priest of the Sun. Heraclitus agrees: "Not I but the world says it: All is one." What makes it one, in his view, is the logos, the precise word of relationship and harmony, adding to a comment he makes earlier: "The Logos is eternal."[4] For those ancients from the West, everything is tied to the earth, our physical and spiritual lives rooted in and of both the earth and the cosmos.

Anglos, too, come from a tradition of humans embedded in Nature, an inextricable part of a single living organism that includes every creature. We have not followed the best of our cultural traditions—no culture ever has for long. But we are not descendants of a cultural tradition that was poisoned at the root as some would have us believe. Indeed, one of the great clues to a satisfactory spiritual life in our time lies in our own culture's most ancient beliefs. Finding the right words—and finding ways to honor their sacred function in the world—is one of the great spiritual and intellectual disciplines for our time.

Clarifying Our Words about Nature

If there is confusion about what it means now to be spiritual, it seems to me that a similar confusion characterizes our talk about Nature and the environment. We have not gotten our rhetoric straight, even among environmentalists, about what we mean when we say "Nature." For most it still means someplace out of town, preferably away from people. We "go back to" Nature. The distinctions we make among Nature, the land, the earth, the environment, the natural world, and wilderness are fuzzy at best. We tend to use the terms as if they were synonymous, but they are not. Worse, our environmental rhetoric continues to betray some confusions in our thinking, and it indicates that we need to take another look at what we say about what we believe. Too often the implication of our words is that we do not—despite our public rhetoric—really believe that humans are part of Nature. Some may dismiss such concerns as mere "semantics," but I think the matter is of graver import. The ghosts of all those I've talked of thus far support me in this.

If we love the world, and if our practice is to be true and our rhetoric revealing of the truth, we have to be clear about our language, and here we are remembering Heraclitus and Confucius, and the Priest of the Sun, and honoring John and Vāc—all of whom honored the word. Yet we environmentalists talk of the natural world as if there were another, unnatural world. We talk of sacred places as if there were other, nonsacred places. We talk of natural materials and plastic, of natural and man-made. We talk of getting back to Nature, as if we could possibly—ever—tear ourselves away. If we humans are inextricably in Nature, linked to it, part of it, as our own environmental rhetoric insists, then we have to forego the language and ideas that imply that somehow we can go to it, as if we were not already in it, wherever we are—mountain height, skid row, radiant coast, New York Bowery flophouse, lush Midwest, downtown Chicago, Sonoran Desert, or the seediest side street of our nation's capital. Yet we seem to think that we can go out and come back; that we'll find something "out there," experience something different "out in Nature" than we find in other places such as offices and back alleys and dank urban canyons, or in our own hearts and minds and actions.

There is no other place; there is only this one entirely natural world, and the trashed-out back alley in the shadows of concrete is as natural as the mountaintop, and as sacred. The roadside strewn with Styrofoam cups and the corner behind the Dumpster covered with vomit from one of our culture's homeless drunks are also natural, and therefore sacred. The red rock arches of the West may be more pleasing, but no more sacred; less changed by human endeavor or human failure, but no more natural. Buber would hold it all sacred. "The holy is not a segregated, isolated sphere of Being, but signifies the realm open to all spheres, in which they can find fulfillment." Buber believed the difference between the sacred and the secular to be fuzzy at best. "There is no not-holy, there is only that which has not yet been hallowed." But hallowing is what we humans do. It is the result of a decision and grows out of commitment. "In reality," writes Buber, "the main purpose of life is to raise everything that is profane to the level of the holy."[5] There is only one sacred place, then, only one Nature in all its distinctive manifestations in the ten thousand things: it is everywhere, and it is up to us in all our frailty, greed, desire, and longing, and with all the self-discipline we can muster, to hallow it all.

This Nature, dependable in its eternity, is our one given in the world. The earth may be blasted, pillaged, deformed, raped, debauched, and despoiled, but Nature is the one thing always, from the beginning, omnipresent, ever there. It is impossible to imagine a world, or a time, without it. It is the ultimate under-around-above-within us thing-being-presence; it is inescapable, eternal, uninvented and uncreated. It is the one being that is all beings and at the same time is a power beyond all beings. However our planetary expression of Nature—"the earth" or "the land"—came to be, from a big bang or a prime mover, from creator or trickster, it clearly came from the still larger, preexisting Nature that encompasses our own little planet as well as all those other planets and galaxies known and not yet known.

There is no way in which Nature and the earth are coequal. We can cause the earth, the land, the environment, great pain; we can wreck its natural beauty and damage everything in it; but it is just another manifestation of our human hubris to think we can change anything fundamental about Nature. The natural processes go on; even at our worst we cannot warp them or shape them to our ends, despite our efforts. Any success in bending Nature to our will is only apparent and surely temporary.

What happens in Nature is a range of mountains, the deep ranges of the sea, the short grass ranges of the American West and the far-reaching ranges of the African veldt, the oak savannahs and tall grass ranges of the upper Midwest, sky-searching midlatitude redwoods and ground-hugging sub-Arctic blueberries, royal elk and dappled fawns and white-face calves, dogwood and poison ivy, big bluestem and Canadian thistles—the incredibly diverse range of species, animal and vegetable and mineral, and the complex range of genders with all their means of pleasure and possibilities of reproduction that support it all. And all of this is not even a tiny fraction of the diversity of the ten thousand ineluctably, deliciously, riotously, profusely, natural "things"—all of them natural, yes, and far beyond our deepest sensibility or power to describe. For us Nature is always both here and beyond. That is, Nature includes not only earth but the Milky Way and Andromeda and all those other galaxies and potential galaxies beyond our most powerful instruments of observation. They are, just as this earthly Nature is, us. If that sounds a bit ethereal or unearthly, even our scientists now say that "generations of heavy stars could have been through their entire life-cycles" before our solar system even formed. The cosmos is an atom recycling bin, and the atoms of those long-dead stars found themselves in the cloud that condensed into our sun. "The earth, and we ourselves," explains the British astronomer Martin Rees, "are the ashes of those ancient stars."[6]

The natural processes go on because there are no supernatural or subnatural or unnatural processes. There is only Nature, of whose fundamental workings we have only the tiniest glimmer of understanding at best. We know so little of it, despite our science, that no matter how closely we associate ourselves with it, we never really know the outcome of any action we take toward it. We pull at a loose thread and unravel a whole ecosystem. We may be able to revision it, revise it, alter it, heal it, change our attitudes and actions toward it, or change our perceptions of it, but we did not invent it and we cannot destroy it. No matter what we might wish, we do not and never will have, thank heaven, such power.

What seems to me beautiful and reassuring about all this is that when the last leaf has strangled and fallen; when the last fish has floated, gills flexing and belly up, to the surface; when the last human has torn off the elaborate clothing designed to protect him from the poisons he has created; and when the earth has blown to smoke like a puffball on the solar wind, what will be left, operating as always in the silence after all our human clamor—and the final whimper—will be Nature. And the rules will still be the same: give in to hubris and Nature will thwart you; abuse it, use it too much for your purposes instead of its own, and Nature will pull the rug out from under you—there will be no place to stand. So the titles *The Death of Nature* (Carolyn Merchant) and *The End of Nature* (Bill McKibben) are wrong. The rest of those books may be wonderfully right, but there is no end to Nature.

We invest our concern and our care in the land, give it our allegiance, and work for its preservation, but we place our confidence, our trust, and our faith in Nature. Whether we destroy it or not, the earth will surely die; that is its nature as well as ours; but Nature goes on. This concept of Nature that I am arguing for is far larger and more encompassing than a single planet. This larger Nature includes us as citizens of the cosmos, participants in decisions of ultimate importance: Will there be human life in the galaxies? What kinds of persons do we have to become to enable a future for these lives (and not only our own human lives) on earth? How do we live to assure our continued presence?

The Pervasiveness of the Sacred

Some people do very nicely without any gods but still feel that there is something more to life and the world than our corporeal bodies and that the life of the world is more than mechanics on a grand scale. They are caught up by the sheer aliveness of it all. They deeply love the world and care for it. Dietrich Bonhoeffer was such a person, a young Lutheran pastor who took part in the famous attempt to kill Hitler. He was profoundly involved in the life of his time and confronted squarely every terrifying moral dilemma humans may face. While in prison, just a few months before Hitler had him strangled with piano wire and hung on a meat hook, Bonhoeffer speculated that the only future for Christianity was to develop a religionless faith that would square with the true nature of the world. God, he believed, was no longer necessary for some.[7] Others remain comfortable identifying a God who created it all and who still cares for all creation—including humans—as part of the creative process of Nature.

Either way, the first great sacrilege is to believe in sacred places, as if there were other, nonsacred places. Perhaps we'll begin to honor best that elemental spiritual life that we seek when we learn to meditate on the darkest corner of the great and dingy city's meanest alley, in the smell of urine and staleness and filth and rotting things well set on their course of decay. They are all natural, and if we love Nature and see ourselves inextricably in it, then our actions are in it too, all our actions, including the creation of desperate circumstances and devastated places. Why would we expect ourselves to be different? Nature, given its own destructive aspects, does the same, and we are in and of Nature. How could we be different? And yet we know ourselves, and we have friends as examples, to be capable of more persistently doing positive, creative work than surrendering to destructive actions.

I am not aware of any scripture, epic, or folktale from any culture that holds that gods reside only in the beautiful. There are many from around the world who hold that the sacred resides in it all, not just its parts. Perhaps we'll begin to honor the gods when we learn to meditate on the darkest aspects of Nature, those that are normally associated with our human actions. That should be possible, for everything that humans do is as natural as anything that a coyote does. The rusting motor block in the junkyard and the dogwood on the edge of the forest are equally natural. Perhaps one is not as aesthetically satisfying as the other, and they play different roles in how the world works, but they are both natural. Therefore, if we love Nature—and see ourselves inextricably in it, as we claim—that back alley scene of a homeless drunk ought to be a source of love rather than antipathy, of acceptance rather than rejection or disdain and disgust, of action rather than impotence. If it is not beautiful or pleasing, then we must work to eliminate both the homelessness and the waste and make it so.

Often what we reveal when we write about Nature is our own desire. We are looking for something: to get away from the city, to find solace or aesthetic pleasure, to be moved emotionally, to learn something. Or we are looking for adventure, for food or water or shelter or resources, for knowledge, for revelation. Whatever it is we seek, we often look to those great and beautiful landscapes we call sacred places for those things, not Nature. Given the state of the world today, how could we possibly focus our sense of the sacred only on beautiful places? That is the landscape equivalent of what Bonhoeffer calls "cheap grace."[8] It is harder, but perhaps more essential, to find the sacred in the ugly, the demeaned, the diminished—the Dumpster that the homeless woman leans against to vomit or pee in a city alley. If you want to see how humans relate to Nature, that is one crucial place to look.

Finding Nature in the marred and the scarred, the ugly and the mean, is harder but also inspiring, for it puts us to work on behalf of a much larger and more inclusive environment than trees and whales, not to protect but to change an environment that includes homelessness and waste and ugliness and disease and injustice. Meditating on that darker environment—the rich and maybe wildest part of all our wild lands as well as a wild part of our own interior life—we find it moves our spirits and minds to create something of beauty where it is most needed rather than where it is most obvious or comes ready-made. When we learn to tell stories about those places and about grand canyons, glacial cirques, and red rock arches, we'll be telling a story that will ultimately create a sustainable life for all the world and all its creatures. This is all part of learning "how to love this world," as Mary Oliver puts it. As much as she loves the out-of-doors, she does not suggest in her poems that we learn to love only benign and beautiful parts of the world but the whole of it.[9]

Our nature writers rarely write about Nature; they write about parts of Nature, especially the parts that appeal to them. Most often they write about landscape, or place, and ignore those elements of Nature that are not beautiful or pleasing. The problem with the nomenclature is that Nature does not reside only in a landscape; it lives in everything, whether the most minute particle or the great vastnesses beyond our globe. In our writing about landscape, we unwittingly dismantle the very unity, disconnect the connections in Nature that we urge others to see. If one wants to look at humans in Nature, one can look at landscape, and one can fret about our impact on it or take refuge in it. But if one wants to witness the way that humans are immanent in Nature, and the way that Nature is immanent in humans, one might also look at the cancer ward in any hospital. When we think of Nature as a pleasing landscape, we may find ourselves with nothing to say to those who are relating to Nature on those other, most intimate terms, where we live with Nature's diseases and where we see Nature as a whole, all-encompassing being in which we live and move and discover—or fail to discover—our own being, and finally die.

Such a view makes some of our claims for landscape sound like platitudes or clichés. "Learn to live with Nature" instead of fighting it or seeing it as an enemy, our writers urge. But what guidance does that give my friends with cancer? Which aspect of Nature do I learn to live with, the part that is trying to kill me or the part that heals my body and feeds my soul? They are both expressions of the same Nature. There, in a larger context that includes disease, is Nature hard at work, the handiwork on proud display. And there are humans hard at work relating to Nature's manifestation as illness. How do we look at that and recognize in those lifescapes inspiration equivalent to that which beautiful canyon landscapes offer us with no effort on our part (except blisters soon healed)? Such friends' hospital rooms and homes and lives are so filled with grief and potential grief, not to mention physical pain—all of it so ineluctably natural in its causes—that one might well come to hate this Nature that seems to destroy us without qualm or care.

So for those who love Nature, one spiritual task for our time is to look at our friends with ravaging diseases directly, full on, in all the implications of their circumstances, and see them too as part of the sacred lifescapes we all love where Nature does her splendid thing. Our nature writing will help us greatly in this territory, for it will show us the connections among the naturalness of disease, the naturalness of urban life, and our beloved natural landscapes. The connections lie in that deeper, more comprehensive Nature that encompasses all of it, in which it all abides, that affects us all just as we affect it in that endless reciprocity in which all the connections come together, the parts are seen whole, and the fundamental unity of all things is made clear.

The Infinite Reciprocity of Acceptance

The universe accepts me; I accept the universe—all of them, however many there are, however they work, the full mystery of it all. I learned something about this on a solo backpacking trip across high plains sagebrush desert from Red Lodge, Montana, to Shoshone, Wyoming—a couple hundred miles of scarce water, heat, wind, and, once, heavy snow that disappeared in a few days and returned the landscape to dust as suddenly as its alchemy had turned it to boot-sticking gumbo. Out there alone, miles from other humans, I felt the full indifference of Nature. For some that experience sends a chill through the spirit. The chill we feel comes from the fear that it is not just this landscape that is indifferent but the universe, all the universes, that as far as we can extend our minds or souls, there is Nature, and it is thick, perhaps even dark, with indifference.

But there is something more. The indifference is real, for nothing out there cares whether you live or die. There is no ill will in Nature, and there is no apparent goodwill either. But within that indifference, beyond the landscape, beyond the stars and within them all too, there is also a kind of acceptance. Purpose is not an issue here. Nothing out there wishes you were other than you are; nothing out there cares if you are just or unjust, wise or foolish, good or evil. My friends, my spouse, my church, my boss, all society may wish me other than I am, but Nature does not. Though you may learn from it, Nature will not insist that you change your life; it takes you, all of you, just as you are and makes no complaint. That acceptance, that absence of judgment, I think, is one unrecognized source of the solace we find in remote and beautiful landscapes. We walk out into the great landscapes to find our fulfillment. But our fulfillment is not in the landscape. Rewarding as the landscape is, we are not fulfilled in it but in that vast, indifferent, utter silence that lies behind it, was there before it, and will still be there after it is gone.

There is in that acceptance a kind of compassion as well. Its absence of judgment includes a kind of forgiveness that we all need. Both acceptance and compassion are prerequisites for love, the love that accepts us entire, as we are. That love can be found at the heart of Nature, that larger Nature that transcends all landscapes, our little earth, even our universe, that lies behind all our cosmos. The universe accepts me; my spiritual task then is to learn to accept (love) the universe with all its powerful goodness and its abhorrent, implacable disease and its ultimate capital punishment. My spiritual task is to tie my spirit to its accepting spirit, so that I can extend that acceptance to all other creatures, every landscape, to all disease and to all health. All of that is natural, all of it deserving of my love because every expression of Nature's voice is, like me, part of Nature, all natural and deserving.

I love Paul's metaphor that describes the church as "many parts, but all one body" (Rom. 12:4–5). But in our time we need to extend the metaphor, for what Paul did not know, had no reason to think of, we do know and must think of: the beloved community includes all creation, not just the human part of it. We cannot take care just of our human needs, or we will not take care of "God's world." There is a principle here, a law. Our compassion has to include it all, out to the farthest star. If it does not, those of us who love "God's creation" will find ourselves at war against that creation. We must love even what is anathema to us and make no judgments on behalf of whatever god we have faith in. Whatever we withhold or exclude from our love (gay folks, straight folks, transgendered folks, white folks, black folks; Ebola, cancer, the common cold, flu) stunts our growth and diminishes our potential to love our God and become God's servants in the only world given us. Extending Paul's metaphor extends our field of view, opens us up to possibilities we may have seen before without the urgency a second look from a new perspective can bring. We may move from "That's interesting" to "Wow, that's really important!" Then we are really open to the possibility of care.

Nature asks nothing of us, but our experience of it does. It asks that we extend the acceptance in the heart of Nature through us to everything else, that we do not insist that the world change to suit our desires or to be what it cannot be. What it cannot be is the infinite resource to serve our economic systems, or a commodity we exchange for personal affluence, or a place where life is easy for us. It can be our inspiration, our solace, an aesthetic pleasure, a balm to our psyche, yes, even a measure of the world's health and our own, for we are healthy only as long as Nature is healthy. But Nature is not here to provide us solace, inspiration, pleasure, or health. It has no intent toward us. We can give back to it the same acceptance and compassion it offers us by having no intent toward it, except the intent that Schweitzer recognized nearly a century ago, the intent to continue to live. That intent, he tells us, is the source of reverence for all life.[10]

A Spirituality for Our Time

WHAT WE YEARN FOR IN the midst of our lives of mostly minor-league chaos and occasional major-league catastrophes is some kind of stability, something solid and dependable, someplace to put that trust in life that we cannot live without. Part of our spiritual and intellectual discipline is to learn to trust the world, to discover that stability, and to keep ourselves together when our world falls apart. We once believed that we could put our trust in the "laws" of Nature; now we are not so sure. Heisenberg sends a shiver through us; relativity makes us wonder; we fear that chaos lurks behind the patterns of our lives.

Our scientists now seem divided on the issue. Once we had a mechanistic universe nailed down and operating according to set laws, purposeful and dependable as a steam engine, if not yet understood. Now there are alternative voices, some scientific, some poetic, saying that Gaia rules again; that the universe is a living organism; that it even has a kind of memory that informs successive generations, a memory so powerful that it can skip generations, coming back after a generation that has never experienced the event that memory recalls; that the cosmos is evolutionary rather than fixed, and that a certain randomness in mutations makes its direction, appearance, outcome all unpredictable, its purpose not merely unknown but nonexistent. Yet there is an irony, perhaps even an oxymoron, in that perceived lack of purpose in our cosmic development. Herman E. Daley quotes Whitehead to make the point clearly: "Scientists animated by the purpose of proving they are purposeless constitute an interesting subject for study." What Daley calls "the lurking inconsistency" in science may come about because "purposeful" and "purposeless" are not comprehensive enough to be useful or accurate descriptive terms for our cosmos.[1]

When it comes to describing the cosmos, we are all trapped in an inadequate vocabulary. One does not have to have either purpose or mechanics, or their alternatives, to characterize our cosmic operating system. It may be more open and comprehensive to say that we live in a cosmos of possibility, one that is not exclusively purposeful or purposeless and that may have some rules (mechanics) that do not change and others (organics) that do. Possibility as our cosmic operating system is more open than either mechanics or chaos, and is of a different order than organics. Possibility is not as vague as it may seem, for it is distinct from chance. Chance needs no prerequisites, has no requirements before it can unleash surprise. Purpose does have requirements to realize its ends. In our human realm, if the requirements are not available, we set out to create them.

Are humans a part of Nature, yet of a different kind—purposeful—than all the rest of Nature? That seems unlikely. Nature is of a piece. Possibility, on the other hand, opens itself to whatever is available. Purposelessness leads to genuine chaos, unlike the apparent chaos that some scientists today say masks pattern. Purpose narrows its focus; possibility opens up. Both purpose and possibility observe, ask questions: "If this swims next to this, what happens?" Both may say, "Ah! Now what would happen if . . . ?" But purpose has an outcome in mind, a hypothesis of hope, a dream of a single possibility. It then begins to narrow down, to increase its focus on smaller and smaller outcomes. Possibility works with what is to create what is not yet. Unlike purpose, possibility may add to possibility without manipulating, without even a hypothesis to prove or disprove, just seeing what may come. Perhaps Nature is more like a child at play than an adult at work. As we have seen, Nature can juxtapose the most unlikely elements to create the most complex relationships, putting the ten thousand things together in ways that are beyond our imagination. As descriptors of the universe, chance seems forlorn; mechanistic seems unreal; possibility seems hopeful, for a cosmos of possibility surely includes us humans as well as everything else. Possibility means we humans may still reach our optimum capacities, may yet become a fruitful and regenerative feature of the world rather than a degenerative liability. In that light a spiritual life that leads us toward sustainability must dwell within us, be available to us, if only those Heraclitean ratios come right to let that possibility blossom.

But possibility can open in any direction, positive or negative. We have seen what negative impacts humans can have on the world. To align ourselves with hopeful possibilities, we can add our personal experience of purposefulness to the possibilities inherent in the cosmos, cultivating our minds toward wisdom, our hearts toward respect, and our spirits toward compassion. It has always been hard to find the dependable in Nature. Purposeful or not, even a cosmos of possibility can hardly be deemed reliable. Many see only total indifference in the wild. We treat the land gently and it ignores us, and our desires, entirely. We do our best to make the land healthy and it comes roaring down in a blizzard that kills our sheep, our calves, or our crops and perhaps even takes our children. Nature knows its own mind, and it does not exist to know or satisfy or even acknowledge ours. This is not a matter of Nature's will. It has no will and no capacity for knowing or acknowledging either our will or our desires.

In his essay "Nature," Emerson holds that Nature is neither reliable nor dependable: "Throughout Nature there is something mocking, something that leads us on but arrives nowhere; keeps no faith with us. All promise outruns the performance. . . . We live in a system of approximation. Every end is prospective of some other end, which is also temporary. . . . We are encamped in Nature, not domesticated. . . . It is the same with all our arts and performances. Our music, our poetry, our language itself are not satisfactions but suggestions."[2] Emerson sounds a bit like a farmer who complains that he did everything right and still didn't get a crop. Or perhaps even more like a contemporary physicist who has worked her way back to the smallest unit of existence she can find and then is confounded, finding another, more elusive mystery she can neither name nor describe. "Quirk" may have been as accurate and useful a name as "quark."

Nature has a heart of mystery for poet Robert Frost as well. In his book about Frost, Richard Poirier asks, "How does anyone know beans? More perplexing still, how does anyone know that he knows them?" In his "poems of work and in his work as a poet," Frost's answer is that "you 'know' a thing and know that you know it only when 'work' begins to yield a language that puts you and something else, like a field, at a point of vibrant intersection." But even then, Frost concludes, "what you finally can know that you know is mysterious and dreamlike."[3] Apparently, neither Frost nor Emerson found Nature a dependable source of either being or knowing.

Martin Buber also insists that at the heart of things there lies mystery. His sense of the mystery differs from Frost's and Emerson's, for this mystery is neither a threat nor undependable. Instead, it makes everyday life sacred—"hallowed" is the word Buber would probably prefer—because everyday life is the place, the arena, the ground (the only ground so far as we know) whereon we confront the mystery and have the opportunity to live with it. That mystery, says Buber, "which is only a gate—and not, as some theologians assert, a dwelling," is the gate through which we step forth into the everyday, "which is henceforth hallowed as the place where we have to live with the Mystery."[4]

As science has come to acknowledge the mystery at the core of things, scientists sound more and more like humanists, theologians, philosophers, and poets. Perhaps the only real difference now is that many scientists are still unwilling to give the mystery a name, while theologians and philosophers—and some poets, like Robinson Jeffers —call it god or even God. But for Kitaro Nishida, an early-twentieth-century Japanese philosopher, coming to know that mystery at the heart of a Nature that goes far behind mere universes or galaxies is "pure experience." For him pure experience lies behind, or before, consciousness; behind, or prior to, time; behind, or prior to, the land or the planet earth. Because it exists at the deepest heart of Nature, it is eternal. We can rely on pure experience, Nishida holds, because it is always available to us, always present. Further, in that depth there resides a unifying force that Nishida calls "spirit." He says, "I contend that reality comes into being through interrelationship," and we hear echoing down the centuries Nishida's ancestors speaking from the Eastern side of philosophical thought, and Heraclitus and others speaking from the West, all testifying to the underlying unity, the harmonizing power and balance that ramify through everything, all reminding us that we can live only in relationship.[5]

Robinson Jeffers, a distinguished American poet in the early 1940s, was also looking for something dependable, that is, a place to find what is eternal. If one could find that, he thought, then one could also find a refuge, and strength to endure life despite the horrors perpetrated by man. For Jeffers our human assault on Nature was so violent that he foresaw a time when only the sunset's colors would persist and humankind would be gone. In poems both long and short, he looked forward to that day. Both Nishida and Jeffers ultimately find that eternal, pure experience only in the totality of all Nature. And for both of them the world is dependable in a way that is fundamental and enduring. But it is not in Nature as we usually define it. Pure experience, the eternal, lies beyond the land, beyond earth, in that Nature that is the pulse and rhythm of the entire universe, indeed lies behind all the universes. Though the earth might disappear, both Nishida and Jeffers would hold, Nature will survive very nicely, and continue.

In "Carmel Point," Jeffers speaks of "the extraordinary patience of things!" His exclamation is caused by "this beautiful place defaced with a crop of suburban houses— / How beautiful when we first beheld it." Then it was "unbroken field of poppy and lupin walled with clean cliffs; / No intrusion but two or three horses pasturing . . . / Now the spoiler has come." He asks of the place, "Does it care?" His own answer is "Not faintly. It has all time." Nature, for Jeffers, is alive, aware, even self-conscious: "It knows the people are a tide / That swells and in time will ebb, and all / Their works dissolve." While it waits for the human tide to recede, "the image of the pristine beauty / Lives in the very grain of the granite, / Safe as the endless ocean that climbs our cliff." Nature has "all time" on its side. Our time, in comparison, will be brief. Very brief, unless we "uncenter our minds from ourselves." The aliveness of the earth, not self-centered but egoless, will simply outwait us. In "Their Beauty Has More Meaning," Jeffers writes about the moon, hanging "low on the ocean," and describes how "the night herons flapping home wore dawn on their wings." His life, like our lives, is short when set against the ongoing life of Nature.

> I know that tomorrow or next year or in twenty years
> I shall not see these things—and it does not matter, it does not hurt;
> They will be here. And when the whole human race
> Has been like me rubbed out, they will still be here: storms,
>   moons and ocean,
> Dawn and the birds. And I say this: their beauty has more
>   meaning
> Than the whole human race and the race of birds.

In another poem Jeffers writes of an old man on horseback riding up into the coastal hills on his way back from a day of ocean fishing. "And nothing," the old man thought, "is not alive . . . I see that all things have souls. / But only god's is immortal. The hills dissolve and are / liquidated; the stars shine themselves dark." The earth will disappear, but Nature will remain. Behind the hills, prior to them and remaining after they are gone is . . . even Jeffers has to give it a name, and the only name that seems appropriate to him is "god." This is where Jeffers seems headed in many of his poems. Though one may argue with the name, Jeffers is on the right track. Nature is, as Nishida insists, our pure experience, the intersection where the knower and the known, the subject and the object, are one, a place of no ego—which is exactly the place that Jeffers is looking for as well: a place that our human ego, being absent, can neither wound nor destroy.[6]

Emerson seems to summarize both Nishida and Jeffers when he remarks, "Many truths arise to us out of the recesses of consciousness. We learn that . . . spirit creates; that behind nature, throughout nature, spirit is present; that spirit is one and not compound; that spirit does not act upon us from without, that is, in space and time, but spiritually, or through ourselves."[7]

Jeffers, though ostracized and dismissed for his "inhumanism," is not entirely alone among poets in his worldview. War colored Sara Teasdale's view of humankind too, and she, like Jeffers, finds both hope and comfort in the end of our human world. Even her phrasing sounds like Jeffers's.

> There will come soft rains and the smell of the ground,
> And swallows circling in their shimmering sound;
>
> And frogs in the pool singing at night,
> And wild plum-trees in tremulous white;
>
> Robins will wear their feathery fire
> Whistling their whims on a low fence-wire;
>
> Not one would mind, neither a bird nor a tree
> If mankind perished utterly;
>
> And Spring herself, when awoke at dawn,
> Would scarcely know that we were gone.[8]

Working Papers on Spiritual Disciplines

I have been trying to build a case here for the connectedness of things: red rock arches, high Cascades, malignant tumors, the dankest city alley. I'm groping for the connections within the outgrowth of Nature, at least metaphorically—image and metaphor being the only tools we have to describe the indescribable mysteries around us—from the logos, the word, and the ties between Nature and our own interior lives, our spiritual lives, if you will. At the same time, I want to keep before us the ties between that spiritual life and subsistence, the ultimate realities that ground us, and sustainability—our current word for a possible future. These are links in a great chain of existence. And I'm trying to make a case for the idea that clarity in the language is one key to the development of a healthy spiritual life. Clarity in the language is also the root of our own personal integrity and will lead us to a worldview that will help us to create a sustainable culture for all. Clarity in our language is more critical for our spirits than any method of prayer, any reading of scripture, or adherence to any creed.

As we examine those links, some may see that language might provide a greater key to a contemporary spiritual life than a just society, a sound economy, or even a healthy environment, for language itself is a "wild" system, as Gary Snyder points out, and another natural element in our human nature.[9] It may take precedence over all those because it will put our lives in a position of integrity, authenticity, and authority that will enable us to heal ourselves and motivate us to work on healing the rest of the world, impelling us to maintain wilderness, create justice, and treat the environment, including our fellow humans, with respect.

I've been trying, then, to think through, and provide a certain coherence for, what I really believe about the world, and yet I still cannot claim to know what all that might be. I wrestle with these ideas all the time, and I have spent a lifetime thus far thinking about these things without, alas, coming to any definitive answers for any of the most serious questions.

So what do I believe about our spiritual life? What constitutes an appropriate spiritual life in our time? I believe those questions indicate that the great question for our lives is not, What am I going to do? even though our educational institutions at every level are aimed at helping us to graduate with skills to do whatever will get us jobs. The prior and more serious question is, Who do I want to be? Howard Thurman, one of twentieth-century America's great mystics, told the members of his Spiritual Disciplines and Resources class that for every person "there is a necessity to establish as securely as possible the lines along which one proposes to live" one's life. He called it a "working paper" that each of us must develop. That, I think, is one guide to an appropriate spiritual life. Our task is to create a life here that works not only for us but also for all life. Some folks set out to have a career in business or law or academe or the arts or agriculture. But I've known others, like Thurman, who set out to create an integrated life, one that joins being and doing in a positive way. Both require a working paper that is expansive enough to allow us to create wholeness, to integrate all the broad range of experience from exaltation to loss, from a sense of life's highest gifts and privileges to the extinction of our lives and the loss of all.

What I recognize in myself is that I have not always known what I was going to do for a livelihood or on behalf of causes I care about, but I have known for a long time what I would like to be. There has been a working paper in place since I fell under Thurman's spell, perhaps even earlier, though I had not the language to recognize it precisely. It remains a working paper because I have fallen far short of the plan much of the time, a fact that makes T. S. Eliot's lines, for me, among the most poignant in all American literature:

> Between the idea
> And the reality
> Between the motion
> And the act
> Falls the shadow.[10]

Nevertheless, the idea of a proper being for myself has remained pretty clear—perhaps never more clear than when, under the light of reality, I have just betrayed it utterly and stand unmasked before a mirror.

The case I'm trying to make has to seem true to me as I examine my own life to discover where I have given my allegiances over the years, what ideas and causes I have tried to uphold, what it is I really believe and want to work for. It also has to include room for both science and spirit, for what I believe has to be based in the best knowledge that comes to hand and built on the best wisdom I can find in others or discover in myself. Spirit for me has come to mean, in part, the amalgam of the best intelligence I can bring to my life and the best heart I can muster. Together, balanced and harmonized, those two become more than the sum of their parts. That whole is what I would call spirit. Others might call it soul. But for me there are two other elements in this:

soul is what in us desires to heal the brokenness, bind the wounds of the world. The spiritual life then includes whatever thoughts and activities lead us toward that goal. For Nishida spirit seems to be the unifying force in Nature. Emerson and Empedocles have the same bent. Breath, word, logos, and vāc have a similar resonance. I know there is mystery in me (would there were only one!) that has a unifying element. I assume it works in others as well. Spirit is a name for that mystery in us that takes all the wounds and all the epiphanies, the anger and the compassion, and keeps them in balance. It is the unifying force that holds us together when we feel ready to fly apart.

To my mind, these notions of balance and unity seem to be quite like the Chinese qi. "In East Asia," explains Mary Evelyn Tucker, "naturalism . . . is characterized by an organic holism and by a dynamic vitalism." The holistic aspect of material force is the enabler "for self-cultivation in harmony with nature." In the Confucianism exemplified by Kaibara Ekken, the eighteenth-century Japanese scientist and philosopher, "the universe is seen as unified, interconnected, and interpenetrating. Everything interacts with and affects everything else." This unification comes because "all life is constituted of Qi, the material force or psychophysical element of the universe. It is the unifying element of the cosmos and constitutes the basis for a profound reciprocity between humans and the natural world." The other aspect inherent in qi is its "dynamic vitalism," which is "the continuing process of change and transformation in the universe" and thus is a key to our own self-cultivation and the development of "an integrated morality."[11]

The cultivation of that spirit so that we become a means to health in a world that, in its inherent unity, combines both health and illness, is the direction I've hoped my own life would take. I do not have regular times for prayer or meditation, and I am not sure exactly what those are or how they work, but I have been aware of times of gratitude and times of desperation so powerful that I felt full as a swollen tick. Perhaps that is one definition of prayer.

However prayer works, for years three ancient prayers have lain in my mind, often dormant for long periods and mostly subliminal. Occasionally, as recently when I was driving among the ubiquitous semis on I-35 north of Austin, Texas, one pops clearly and apparently unprompted into the forefront of my mind. It is always one of two verses from Psalms, as it was the other night on the road: "Create in me a clean heart, O God" (Ps. 51:10). Experience tells me that is not going to happen, is not possible.

But experience also tells me that one step I can take toward that end lies in my language. The clearer, more direct and honest my language becomes—without causing gratuitous pain in others—the clearer my mind becomes, and the cleaner my heart seems to come. Language is not only the means to good government, as Confucius says, but also one means to a healthy spiritual life. When I get my words straight, I increase my capacity to get my life straight, straightened out. So creating in myself a clean heart is more than routing out simple lust; it involves straightening out my attitude toward the world and others, dealing with the anger, disappointment, loss, and pain that inevitably come into every life and that occasionally threaten to sink mine.

On balance, given the life I've led, so enormously privileged compared to those of most people in the world, I have to acknowledge that I have no right to my anger, that the annoyances that give it rise are mostly just that—annoyances, and petty. I've only had one job, ever, that I didn't like. My work has achieved some good ends on occasion, and I've received some recognition from it. I have known men whose example has drawn me to a harder-working, smarter, more thoughtful life than I could ever have managed on my own. I've been lucky to love a number of women and can say that I still love them without being misunderstood by any of them, including my wife. My children are still speaking to me, and my pride and pleasure in them waxes daily. My brother means more to me every year. I have seen indescribable beauty in the natural world and have tasted its threats too: Once my first wife, Shirley, and I crouched near the stove in the center of our tiny house in Iowa while a tornado, passing less than a quarter mile away, blasted all the doors wide open and scattered hail into the corners of every room, and we wondered if the whole place would collapse. In Montana and

Alaska I've walked (always unwitting, always astonished) within a few yards of bears both black and grizzly.

Once, fishing alone, I was wading Paul's Creek in Alaska. It was late fall, the day overcast, the tidal river running on the rising tide, moving silver salmon up the creek. I was hoping for silvers. As I was casting ahead, watching the water closely, approaching a small island that divided the river, my attention was diverted by a movement in the dry, beige beach grass that covered my end of the island. As I looked again, an Alaska brown bear rose slowly upright, trying to focus his poor eyesight on whatever was encroaching on his territory. That bear looked huge to me as I stood thigh deep in the tidal flow. Some Alaskan folks say that one good thing when faced with a grizzly that size (or any size) is not to appear aggressive. That was easy. I did not *feel* aggressive. Standing in the river, armed with a fishing rod, water almost to the top of my hip boots, I knew in my bones the idea that humans are at the top of the food chain was nuts. I was not about to lose my fishing gear, or a leg, if I could help it, so I talked softly to the bear as I started to back away. "Sorry, Bear. Yes, I know this is your territory. You can have it; yes, I'm backing off here, Bear. Sorry to intrude. It's all yours, Bear," backing down the creek, Mr. Obsequious himself, reeling in line, trying to keep from falling, trying to keep the water out of my boots, trying to get to shore, bowing, talking to Bear.

So I've not only experienced the aesthetic epiphanies that great landscapes have to offer, but I've known fear. The biopsies I've had and the infrequent but powerful times when my body has had mechanical failures remind me that there are other places, far more intimate than great landscapes, where Nature operates, that human life absolutely is immanent in Nature and Nature is immanent in us, and the thinking of Newton, Bacon, Descartes, and Kant never changed that for an instant. They may have warped our perception of that—given too much rein to the hubris that would like to think it is possible—but they never altered the reality one whit. Man thinks he can sever that tie, Heraclitus muses, but the thought is clearly foolish. Given all that, how could I be anything but grateful?

And yet, even amid the general gratitude, there are times still when anger boils up, or disappointment saps my spirit, or fear overtakes me, or loss overwhelms me with feelings so powerful I do not know what to do. I look at what is happening to America's growing numbers of poor and listen to Congress talking about it—and I hear the economists telling us about the strength of the economy. I think about the world's poverty and how my own life helps to perpetuate it, I watch the ease with which my country destroys those who do not do its bidding in the world with potent weapons, and I see great landscapes laid barren for the wealth of a few, and I get mighty irate and depressed. Friends now dead or dying, the loss and the potential loss rise like bile in my throat, bring tears to my eyes. All of those feelings are natural, and I want to acknowledge all Nature and accept it gratefully in all its manifestations, yes, but I also want to align myself with the forces in Nature that strengthen and heal and regenerate. Making my heart "clean" then means there is still much work to be done. Despite my Christian upbringing, I do not believe any god is going to do that work for me or take me off the hook; the forces in Nature that I want to work with require cultivation within me, undertaken by me, with whatever additional resources I can bring to bear.

The second prayer also comes from Psalms. My attention was drawn to it as Thurman began each Sunday's sermon with it: "May the words of my mouth and the meditations of my heart be found acceptable in thy sight, O Lord my strength and my redeemer" (Ps. 19:14). There is a clear tie between the former prayer and the latter; they always come to me in tandem, never one without the other. Both, for me, have to do with language; both make a direct link between our words and a truer life, as if the closer my words lie to the word, the closer I come to rectifying my life through the alchemy of linguistics, the power of language. That "truer life" does not necessarily mean a more moral life, at least not one that is moral enough that others would notice. But it does foster the possibility of a life with greater internal coherence, as Confucius knew, a life of greater integrity, greater allegiance to those forces in Nature that generate peace rather than pain and health rather than anger, and a life in which my perceptions are clearer, more real, more authentic. My own capacity to delude myself is then reduced, and I too have a greater chance to become more authentic, more real, more balanced and harmonized, more a self I want to think is my true self.

I do not remember a time when the third prayer did not come with the other two, all three together. Perhaps it does not come unbidden to my mind; perhaps the other two carry it with them. This third prayer has the power, for me, of extending my self-concern and self-cultivation to the wider world of society and landscape. Any spirituality that ignores either the local or the global culture, that does not aim toward rectifying our human society and all those related societies we now call ecosystems, at the same time we rectify our hearts is a false spirituality. I heard this prayer from Edwin Prince Booth, a history professor at Boston University's School of Theology. He used it more than once on public occasions. I never learned it, really, but I never forgot it either: "Grant unto us, O God, such a vision of thy being and beauty that we may do thy work without haste and without rest." If one comes to see Nature as the means we have to perceive the being and beauty of God, if Nature and God are inextricably linked, if the word is the instrument Nature uses to hold all its various elements in their rightful place, and if Nature and the word are cocreators, perhaps akin to *qi* in Chinese thought, then that vision both drives us to beautify the worlds accessible to us and sustains us in that effort. For me this prayer asks us to do what Nature asks us to do: to extend or expand the acceptance and compassion at the heart of the cosmos to everything else, to make our spiritual life active in the world.

I do not mean to imply that because I carry these prayers around in me that I am a good person or a religious person. I mean only that they seem to steep in my mind someplace mostly subconscious, and I suspect that without their occasional rising to consciousness I would be a different person. When they do come to mind, they force me to look at myself, and in the looking it usually seems that they have come to mind for a reason: I need to be reminded of them because I am screwing up somewhere inside, or my will to work on my self and for the world is flagging. They exert a discipline on my life, a call to examine my own conduct rather than the conduct of others. They are also a resource that helps sustain me. Sometimes, if I am alone, I find myself surprised, saying one of these prayers aloud, throwing the sound out into the darkness like a life ring, hoping it will save me.

Self-cultivation is thus a spiritual task, not simply a psychological or social necessity, and it requires discipline; indeed, in his class Thurman also told us that the disciplines he had in mind are not undertaken on behalf of the self or for the sake of the self; neither are they a matter of chalking up points toward salvation. The purpose of spiritual discipline for Thurman, as for Confucius, is to cultivate a self as whole and fully human as possible so that we create a reservoir, a well, something sustaining and of substance underneath or beyond the surface, something more than mere appearance. Out of that reservoir we can draw on the self to give of ourselves to others and to the culture: respect, attention, care for the other—whether fellow human, other species, or others' ideas or thoughts. We will consider our prejudices; our role in the continuation of poverty; our effort to recreate, restore, or regenerate an earth under siege. We will train ourselves to be an instrument of Nature—that Nature which knows that everything needs nourishment for body, mind, spirit—knowing that our ideas and our faith are connected to our actions. We will pay attention and offer our reverence for it all. We strive, then, to make our self of a piece, just as Nature is all of a piece, everything connected in us, contributing to the whole, an ecosystem of the soul. We will pay attention not only to our personal and civic actions, and to those people we meet in the course of our lives, but also to how we live in the landscape, to those creatures who, Schweitzer says, all share our will to live and are therefore kin to us. That is how we train ourselves to be and do in a world we will transform in a positive way. It is also the intellectual, spiritual, emotional task I described earlier, and it is the kind of self-cultivation that will lead us toward Prescott Bergh's sustainable culture.

We are not here, as some believe, simply to discover who we are or to fulfill our own nature. Rather we are here first to fulfill the nature of Nature. When we work on that, we fulfill ourselves in the process. That is the work Buber outlined for us earlier, and it is a matter of being in the world in such a way as to bring our deeds and our words together, putting, as Heraclitus said, our "acts and our words to the test."[12]

In 1708, at the age of seventy-nine, Kaibara Ekken published *Yamato zokkun* (Precepts for Daily Life in Japan). Mary Evelyn Tucker, whose grace- ful translation of this work and whose insights into Ekken's life offer us a compelling and thoughtful introduction to his writing, tells us that Ekken was first an important scientist, a biologist and botanist, whose books, articles, and drawings were careful studies of the plants, fish, and birds of Japan. His love of Nature and his study of the classical Chinese texts, especially the works of Confucius and his followers, fit together, and one might say that his "ethicoreligious" stance toward the world grew directly out of both. Nature was both source and pattern for his self-cultivation and his actions in daily life. Further, his classic Confucianist view of family as the great metaphor for all our relationships is as appropriate for us, and for our educational systems, as it was for early-eighteenth-century Japan: "Humans have heaven as their father and earth as their mother and receive their great kindness. Because of this, always to serve heaven and earth is the Human Way. What is the Way by which we should serve heaven and earth? Humans have a heart of heaven and earth, namely the heart of compassion which gives birth to and nurtures all things. This heart is called humaneness. Humaneness is the original nature implanted by heaven in the human heart."[13]

Reflecting that endless reciprocal compassion between earth and all its creatures, Ekken insists that every ethical attitude is rooted in our gifts from Nature. Because we are children of Nature (heaven and earth), we have "a heart of compassion"; because we have a heart of compassion, we have a responsibility to turn that compassion alive and loose in the world. "The principle of humaneness makes it a virtue to show kindness toward human beings and compassion for all things," Ekken says. He explains, "The way to serve heaven and earth is by preserving this virtue of humaneness without losing it, and deeply loving humanity, which heaven and earth have produced. Then by having compassion for birds and beasts, trees and plants, and adhering to the heart of nature through which heaven and earth love humans and all things, we assist the efforts of the great compassion of heaven and earth and make the service of heaven and earth the Way."[14]

For Ekken, living with humaneness in the world is a matter of study and of spiritual practice. The study he upholds for all, whether of noble or peasant birth, is the study of the great classics. But study, to be worthy, must reveal itself in action: "There is no other way to serve heaven and earth than to obey the mind-and-heart of the universe. Obeying this heart . . . implies extending warmth, compassion and respect." Humans are born and "through the beneficence of nature, they receive its heart and make it their own. They live amidst nature and partake of its nourishment." And here Ekken, too, sounds like Heraclitus, complaining that people hear the logos but do not understand it: "Thus they receive an infinitely great favor, but most people do not realize it." To not serve Nature is "to receive the great favor of heaven and earth and at the same time to act contrary to nature." Such a life is "extremely unfilial," Ekken says with great force, speaking from a culture in which to be unfilial is a grave error.[15]

Writing about Koyukons in Alaska, Richard Nelson examines the combination of the natural and the spiritual that characterizes the Koyukon view of Nature. Summing up, he says, "In the moral system that this ideology encompasses, the proper role of humankind is to serve a dominant nature." It is also a matter of recognizing, and acting on, the relationship we have with all creatures, learning to hear the world around us and the songs the cosmos sings through "the music of the spheres," to feel its whirling movement as everything in the world moves around us, to learn what the river says—and say it ourselves.[16]

Once, driving into the growing light of dawn around Turnagain Arm in Alaska, I felt that movement in a visceral way. First the dark and a sense of presences, of things mysterious and huge, looming behind the dark. I knew what lay back there because I had been down this stretch of road many times before. But then the first glimpses: a narrow gleam, a silver sheen on a narrow strip of water, the light gray sand emerging from the tide in its retreat, then mountains looming, clouds lightening, and gradually the pale glaciers, forests bending to the wind, all moving, all coming into the light, all directly connected to me, to all of us, and everything moving, farther than eye could see or mind could fully grasp, beyond stars, beyond galaxies. Now I think Eliot was wrong: There is no "still point at the center of the world's turning," as he claimed.[17] There is only the circling movement, the dancing swirl of great Nature within its infinite and eternal cosmos. The word, the dance and the song, the beauty and mystery of our habitation, and each of us aligning ourselves with the creative powers of the word and of Nature, using our words to tell the stories that will heal us: this is our movement and our song, the reason for our being. If we can give ourselves to such attention to the world, we will take our place within a system Koyukon and other peoples have known for thousands of years.

Seyyed Hossein Nasr traces the development of "the Order of Nature," a schema for the world that allows humans to see order in chaos, from its earliest, indigenous beginnings through the development of the world's great religions and their view of a divine order in Nature. What Nasr finds beguiling in those views is the similarity of their structure. For the most part this order was designed by the structural cultures that I discussed in an earlier chapter. But once again it is amazing that people of so many different cultures and times develop such common views from their disparate environments, traditions, and perceptions. That order of Nature does not now sit well with contemporary science, but the problem with it is not that it lacks scientific rigor but that it is our human fear, awe, desire to understand, and our perception of the world imposed on Nature. We order it, whether it is ordered or not, and in our hope, fear, and arrogance, we put ourselves at the earthly top of the ten thousand things—a little lower than the angels, we concede, our humility more than a little lower than our arrogance.

But what if we reverse the process and let Nature impose itself on our worldview? Nature has always been our teacher, but in stories and lore in the West it has also been bent to our desire for order. Our worldview of the order of Nature stems less from what we have learned from Nature, more from our hope that Nature is ultimately ours to use, and if we can only find the order in it, we can bend it to our will, establish some control over it. Forget order, our culture finally holds. Go for control. In that one move we defy the entire operating system inherent in cosmic possibility, closing off potential futures for ourselves.

But we can let Nature provide the means to our worldview, a world-view that can evolve (and has evolved, whether we acknowledge it or not) as Nature evolves, in fits and starts and random discoveries and gradual growth and development as the possibilities emerge. Then we learn to listen to the river, as William Stafford says, and learn to echo the river in our thoughts so that, finally, "What the river says, that is what I say." Then Nature's innate self, ordered or not, becomes our self, our thought, our means for looking at the world. Rather than looking for a pattern in Nature we cultivate ourselves to fit Nature, discernible pattern or no, and allow ourselves, as my son Kevin says, to become both natural and truly civilized.

From the Beginning

From the beginning there has been a word between us. In the nature of things we say a word to another. It fills everything in, closing the gaps between us, creating a bond. There is also a word between ourselves and all other creatures, ourselves and all Nature, a word even between ourselves and that ultimate mystery we hesitate or refuse to name. Heraclitus and John and our oldest texts, Eastern or Western, indigenous or not, oral or written, all were right in this: that word, the name we are reluctant to give to the mystery, is what holds everything together. If we don't keep up our end of the conversation, things fall apart. If we do keep our word, then the creative power of Nature and the logos, vāc, and the world's breath is realized through us in the world. One sacred means to achieve harmony and balance, to be the chord that sounds between humans and everything else, is simply the clear word. I have said it before: the spiritual quest for us is, like Abel's in *House Made of Dawn*, a search for the right words.

Once again, a remarkable model for this comes from another culture and an earlier age. Yi I was a Korean philosopher and government official in the mid-sixteenth century known by the pen name Yulgok. For Yulgok the cultivation of the self, the very possibility of learning, begins with a focused will. That focus comes from sincerity, a notion that he picked up from Confucius and his followers. Yulgok, writes author Young-chan Ro, held that "the process of becoming sincere was in large part a linguistic one, the disciplining of one's use of language." For Yulgok, then, "language can become much more than communication. It can become a mode of being and action, and when language becomes the true manifestation of both being and action, it becomes sincere." Yulgok himself wrote, "Learners . . . must be careful regarding language. Many human faults come from the misuse of language; so words must be careful, truthful, spoken at the proper time, and given in a serious manner." Ro stresses that "this process was not in any way metaphorical for Yulgok. . . . Being is dependent on language, but language must be combined with sincerity." When that happens, words and actions "gain the power to shape and define reality." Indeed only the combination of thought and action, sincerity and language, can create what Yulgok calls a "sincere reality."[18]

What I can say about our spiritual life in these days is that we have to create our own. The only spiritual life that will work for us is the one we work at creating ourselves. No institution can do it for us. Humans, like all the ten thousand things, are sacred. Our institutions, even our religious institutions, are not. Our religious institutions are no more sacred than our military or our government. It is but idolatry to believe so. We may give our allegiance to a particular faith or church, follow its disciplines and practices, even believe what it tells us to, but until that faith is tested in the crucible of our own experience, it is not truly ours. Always our experience tests our faith, and we create our faith as we work at our own lives, live out our own experience, adapting the elements of faith that work, that remain believable for us, and ignoring or sloughing the rest.

Further, no one can easily appropriate a spiritual life from other cultures: if we are not Lakota, we cannot simply steal a glimpse of Lakota spirituality and make it our own. We get something greater if, like Gary Snyder, we immerse ourselves in the ways of another culture for years and live what we learned for the rest of our lives, exploring the meaning until it is our own. Snyder's Buddhism is not that of his earliest teachers, nor that of his latest. It has become his own. It *is* Gary Snyder. We may use what is enabling for us from many sources, including institutions of religion and spiritual resources from other cultures, but ultimately the faith that sees us through is the one that we have accumulated, nourished, fostered, come to terms with—seen to be true in our own experience—and that we can continue to shape as circumstances and events change or even assault us. We grow into that inner life just as we grow into our ideals as we grow older, pulling together the various fragments from many sources to create a coherent whole.

Realizing God

For a time in the 1960s and 1970s, some theologians, echoing Nietzsche, were saying, "God is dead." Buber instead said, "God is silent" in our time.[19] It is up to us, then, to utter, however weakly, the creative word that the Priest of the Sun said was there before the silence and was *in* the silence. In some traditions God sends his emissary into the world. In the Christian tradition God decides to become incarnate; the word becomes flesh. But there is another view that holds that if God is to be present in the world, we must bring God here through our own actions. What Buber seems to suggest is that those who believe in God must bring God into the world themselves, that God is "to be realized through us." But what Buddhists, Taoists, and Confucianists have shown us irrefutably is that one can achieve spiritual strength, authenticity, morality, and an expanding circle of care for humans and the other ten thousand things of Nature without belief in a transcendent god. Further, in acts of huge generosity, they open the doors of participation in their practice without asking others—Christians, Muslims, Jews—to forsake traditional faith in order to join them in spiritual discipline, practice, and praise.

As later Confucianists developed their thought, they saw that we humans stand between heaven and earth and are created out of both. Our human role is to realize the heavenly in the earthly. They were not alone in this view, anymore than Jeffers was alone in his. Heaven cannot realize itself fully without our help. To Munetada Kurozumi, the ties among human and heaven and earth are so obviously immanent that "this heaven and this earth represent the self."[20] If that is the case, the stakes are clearly as high for us as they were for John. One of our culture's healing stories, told by Norwegian novelist Johan Bojer, seems to echo Buber's idea of how God comes to be present in the world and also speaks to the Confucianists' task of realizing the principle of heaven in this world.

Bojer has Per Troen in his book *The Great Hunger* rise to the heights of fame as an engineer.[21] He is invited to the palaces of kings and sought for all over the world as a great builder in steel and concrete. But at the height of his career, things begin to go wrong—a mistake in calculations, workmen killed. His career begins to decline, his health begins to fade, and his life falls apart. Per is forced to go back to his Norwegian homeland and settle in a tiny peasant village where he is surrounded by the uneducated and the poor. His next-door neighbor is a suspicious man who hates everything he cannot understand, and one of the things he cannot understand is Per. The bitterness between the two men climaxes when the neighbor's dog attacks and kills Per's little daughter.

One night after that tragedy, Per sits looking out the window, and he begins a letter to a friend. He has come to realize, he writes, "that great sorrow leads us farther and farther out on the promontory of existence." He has come to the outermost point now—there is no more. He writes to his friend,

> I sat alone on the promontory of existence, with the sun and the stars gone out, ice-cold emptiness above me, about me, and in me, on every side. But then, my friend, by degrees it dawned upon me that there was still something left. There was one little indomitable spark in me, that began to glow all by itself—it was as if I were lifted back to the first day of existence, and an eternal will rose up in me and said, "Let there be light!" [The Word sounds again from the darkness and silence.] I began to feel an unspeakable compassion for all men upon earth, and yet in the last resort I was proud that I was one of them. . . .

There was a drought then. When people finally dared to plant their grain, the frost came hard and late and the seed froze in the earth. The neighbor had his patch of ground sown with barley—but now it was gone and he had no more seed. Indeed there was no more seed to be had. He went from farm to farm begging for some, but people hated the sight of him and his great dog; no one would lend him any, and the boys on the road hooted after him.

One night, as Per lay sleepless, he got up when the clock struck two.

His wife rolled over and asked him where he was going. "I want to see if we haven't a half-bushel of barley left," Per said.

"Why? In the middle of the night . . . ?" his wife asked.

"I want to sow the neighbor's plot with it," Per replied, "and it is best to do it now so that no one will know it was me . . ." He went out into the soft night air. The farms were still asleep. He took the grain in a basket, climbed over the fence between him and his neighbor, and began to sow. From the heart of sorrow, in that letter he wrote to his friend, Per had said, "Mankind must take heed that the godlike does not die. The spark of eternity was once more aglow within me, and said, 'Let there be light!' Therefore I went out and sowed my grain in my enemy's field that God might exist. And when the grain was sown and I went back, the sun was glancing over the shoulder of the hill. There by the fence stood my wife, looking at me. She had drawn a kerchief over her head in the fashion of the peasant women, so that her face was in shadow; but she smiled at me."

Word as God, Word as Nature

In our founding desert traditions, the metaphors we use to characterize God are similar in each religion: "He," "omniscient," "omnipotent," "eternal," "ever present," and "loving and caring for 'His' creation" are general characteristics that fit the God shared by Judaism, Christianity, and Islam. Our differences are less over the concept of God than over who God's messenger, messiah, or prophets might be. But if the word is God and the word is the binding force of relationship and harmony in Nature, if God as word and Nature as word are conjoined, as so many traditions indicate, then the words we use to characterize God are equally applicable to Nature, for Nature exhibits many of the characteristics attributed to the traditional transcendent God of our desert-born cultures.

One of those descriptors, possibly two, should be shed by all religions. One is "he." Changing "he" to "she" is not a help here. Nature is, quite naturally, both male and female and the full range in between. Moreover, in the myriad reproductive capacities of its cells, plants, and animals, Nature is heterosexual, homosexual, bisexual, and asexual. The second questionable descriptor is "omniscient." That word has been twisted over Christian history to mean "foreknowledge," to suggest that God knows everything that is to happen. For some it means that "he" knows beforehand that some of the persons whom "he" creates are destined to suffer throughout eternity, whereas others are to be saved for a glorious afterlife. "He," in this line of thought, can even identify the members of each group ahead of time. Such a god may know more than we do but is so brutal "he" may be worthy of fear but not of awe, and not of worship. Such a god forecloses on two possibilities: the possibility of growth and the possibility of redemption. But Nature operates on a system of possibility, of openness, of acceptance, of relationships so complex and marvelous, so intricate and lovely as to instill awe wherever we direct our gaze.

What Nature asks of us in exchange for its openness and acceptance is our moral responsiveness to whatever conditions in which we find ourselves. One mode of responsiveness is respect—that we respect all Nature including humans—another is being open to possibility in Nature, including possibilities for other humans and ourselves. Therefore, we have an obligation to "grow in wisdom" and not just in stature or in knowledge, to steal a biblical phrase. Wisdom includes growing not only in our knowledge of how the world works, in our self-knowledge and knowledge of our human nature, but in moral wisdom as well. If we are to create a sustainable culture, growing in moral wisdom is an ever more inclusive concern for "the least of these," whether the least be the smallest, least attractive units of Nature or the humans who seem to be lowest in our social world's estimation. Sustainability requires this regardless of our faith. Nature has given us our intelligence. We may argue over whether or not it is as great as the frog's, who is wise enough not to drink up the pond in which he lives, while it appears that we lack such constraint. But we cannot argue over whether or not we need to cultivate ourselves to continue to grow—not only in knowledge but also in the wisdom that leads to a more profound, inclusive, and responsive morality. To grow in wisdom is the natural course for us to take.

Because possibility cuts in both negative and positive ways, we may lose our natural course and quit learning. There are also some natural reasons for us to fail to grow: our capacities may be limited; our circumstances may circumscribe our possibilities. We may have to spend so much time just trying to survive that expediency short-circuits opportunities to acquire wisdom. We may have to work such long hours that we exhaust both ourselves and our time for reflection. We may have had the desire for wisdom beaten out of us by abusive parents. Nevertheless, most of us in this privileged nation, this privileged culture, have little or no excuse for not being responsive to Nature, to one another, and to ourselves. We have had some terrifying models in our time of how our human morality can become crabbed and even vicious and yet parade under winsome, though deliberately false, arguments. But once again, it is the testimony of many cultures, including our own, that we can cultivate ourselves toward a wisdom that is without malice and reveals "charity for all," as Abe Lincoln and King James have it. We can be sure that unless we cultivate a charitable spirit within ourselves, it will never return. We have too many institutions now bent on harsh judgment rather than generosity. Judgment is a religious matter, not a spiritual practice. If we are to err in our spirituality, let us err on the side of generosity rather than judgment.

Nature is also eternal, regardless of its operating system, whether it is mechanistic or living, evolutionary or driven by law, purposeful or possible, as Nishida and Heraclitus and Jeffers all suggest. Nature has ever been present and will be, ever emerging, open to possibility, unfinished, and undiminished, after our humankind is long gone and all the stars we see have burned out. Nature is the one eternity we may be absolutely certain of. Some religions of both East and West fret about our entering into eternal life. Generally in the West, that means we may, by means either of our profession of faith or of our benevolent actions, earn the right to enter an afterlife such as heaven. In the East, it may mean melding into eternity after a succession of rebirths in which we improve our character, shed our human pettiness, and purify and repurify our souls until we are fit to flow back into eternity. But as part of eternal Nature, we humans need never worry about entering eternal life. We are already part of it.

As part of an eternal, ever emerging Nature, open to possibility, we are thrust into this earthly natural existence by Nature, and Nature continues to hold us in this life even as we move to this life's end and join another aspect of eternal Nature without pause. Death is not an event but part of a process. In Nature we are participants in eternity entirely independent of faith and action, holders of a creed or utterly without one, with or without our will. We are always Nature's, in one form or another, accepted without reservation, held as we are, whatever we are, in the presence of Nature. Thus we come from eternity, live in eternity here on earth, and continue in eternity as part of the great cosmic Nature that comprehends all life. Nature does this without ever asking us what we believe or what we have done.

In *The Sign of Jonas*, Trappist monk Thomas Merton quotes scripture: "In Him all things are made and in Him all things exist." We do not injure either fact or faith to say, "In Nature all things are made and in Nature all things exist." Our desert religions insist that we live and move and have our being in God. It is a matter of faith that we live in God's presence and that God's spirit resides in us; it is a matter of fact that we live within Nature and that Nature lives within us. We live and move and have our being in Nature, or as Thomas Berry puts it, "The human is less a being on the earth or in the universe than a dimension of the earth and indeed of the universe itself." It is not necessary then to "realize God through us," as Buber asks. Our task is to realize our own nature and all the rest of Nature through us. Nishida offers a kind of summary that incorporates all the above notions: "There is nothing that is not a manifestation of God."[22] As I read it, of course, he could as well have said, "There is nothing that is not a manifestation of Nature."

Further, Nature is all-powerful, as we know in our personal experience; for all our technological exploits, human power has no strength to withstand Nature's power. Our strongest houses succumb to tornado or quake, and floods—often created by our own foolish land use—sweep our lives away before our eyes. Volcanic ash buries even our most beautiful cities and clogs our engines, all our technology, no matter how advanced their technical perfections.

And that loving and caring for its creation? I find it in that acceptance, compassion, and forgiveness that I described earlier. It comes as we look behind the appearance of Nature's indifference and discern Nature's acceptance at the heart of the heart of the cosmos.

This may seem like old-fashioned "nature worship," now abhorred by many as primitive, animist, or even "savage." If that is the case, so be it. Our culture is as fiercely superstitious about the economy and political and military power as any indigenous culture is about Ground Squirrel and Bear. I will trade in my culture's worldview filled with the superstitions that demonstrate our respect for accumulated goods, wealth, and power above all else. The superstitions of an indigenous worldview that respects all life and engages in rigorous self-cultivation and study, the goals of which are wisdom and survival of the people, seem like a good swap. The former choice—putting my faith in our current system—means I have a ticket on another *Titanic*, soon to go down in its own selfish acquisitiveness. The latter choice means I sign on for a rich, enlightened, and all-encompassing spiritual life that may lead us on a path toward sustainability.

I don't know much about worship, but I admit to awe at Nature's complexity and mystery and to respect for its power and pervasiveness. And I admit to finding solace in its inescapable, all-encompassing character, its indifference, and the comforting acceptance that lies behind that indifference. And, finally, I admit that I am challenged by the effort, never fully realized, to understand it. If these are also aspects of worship, so be that too. To betray Nature is to betray the very "ground of our being," to assault the sacred. If we have forgotten this, perhaps a Koyukon elder will remind us: "The country knows. If you do wrong things to it, the whole country knows. I guess everything is connected together somehow, under the ground."[23]

Whether you call that power God, or whether you call it Nature with a capital *N*, as I have come to do, our task is the same: to put our lives together in such a way that our lives and our words stand together with that word which is in and of Nature, whether it be logos or *qi*. That is basically a spiritual task. Ask Per Troen or the Priest of the Sun. Ask Epictetus or Confucius. At the same time we work for a sustainable life for the earth and for our social world, we have to work on ourselves, on setting ourselves right. This is now, also, the common parlance of our psychologists and counselors: "The only behavior we control is our own." Would that we could manage even that! In that task lies the self-cultivation envisioned by Confucius and his followers, and it is an essential task, one that is never ending, never fully achieved. Somewhere in that notion lies the meaning of integrity, the integrated life, and the key to a sustainable life for all. Ask Confucius. As usual, he said it better two and a half millennia ago: "Only those who are their absolute true selves in the world can fulfill their own nature" (this is that spiritual life we've just been thinking about); "only those who fulfill their own nature can fulfill the nature of others" (there's the social life, the purpose of which is to help others realize themselves); "only those who fulfill the nature of others can fulfill the nature of things" (I'll cheerfully admit that "things" is a pretty vague term, but I choose to believe that it has to do with all those "ten thousand things" around us in the environment); and "those who fulfill the nature of things are worthy to help Mother Nature in growing and sustaining life" (there's the whole of it all put together).[24]

The Chinese character for living up to one's word is made up of two separate characters placed together. I heard this from Sam Hamill almost twenty-five years ago, and I have been reading it over and over, first in Ezra Pound, then in the Ernest Fenollosa he learned from, and since then in many others. This is the character for "human":

This is the character for "word":

Together, they represent *hsin*, a person standing by his word:

What is crucial here is that the word we strive to stand by is not just our personal word. That's just ethics, and ethics don't interest me much. Some call for a "new" land ethic. I think we do not need any more ethics. They are too often a means to separate us: "we" have them; "they" don't. What we mostly do with the ethics we already have is ignore them. We have known for millennia how we should treat the earth. Our problem is not that we lack ethics but that we fail to act on the ones we have. So here I am after something that is deeper than ethics, something not only of the law but of the spirit. What interests me is whatever it is that comes before ethics and prompts us to create them. What is that impulse? Where does it come from? That something, I believe, is the universal logos/vāc/breath, the word that is the ground of an appropriate spiritual life for us all. It leads us to respect and to that natural acceptance of the cosmos as it is, which is the essential condition for love.

We accept it all, the cosmos as it really is, and we love it. But we align ourselves with and struggle for the healing and wholeness of the healthy aspects of Nature. In the process of developing the compassion of heaven and earth in ourselves and extending it to all other creatures, we help Nature realize its own best nature. When we align ourselves with that word, we align ourselves with all the creative natural forces for balance and harmony in the world and focus our creative minds and hearts on the world's present and future.

I've been trying for that all my life—only off and on, I must confess—and have tried to speak the word I believe to be true, though I have not come close to accomplishing that yet. Still, it seems a project worth pursuing and remains the work that is most exciting, for I am convinced that the key to taking care of the earth and the society lies inside each of us. If, as the rhetoric of sustainability insists, the health of humans is utterly dependent upon the health of the earth, so, too, the health of creatures of every kind, and landscapes everywhere, now hinges upon the health of our spiritual life. I have been trying to show that the real roots of a healthy spirituality lie in two arenas. One is our language, rather than any scripture, creed, or institutionally transmitted belief. The other is in our self-cultivation toward humaneness.

Though our human nature is "natural," David Armstrong, a thoughtful biologist at the University of Colorado, reminded me that it is also peculiar to us, somewhat different from the nature of other species. That peculiarity, David said, has allowed us, alone among species, to burn the earth's energy faster than Nature can replace it. Now the earth has grown so crowded with our ubiquitous presence that we wear it down and wear it out. We can affirm what Lucretius believed in the first century, that the earth is getting old. But when we get our most interior, spiritual lives in line with the eternal logos (or the cosmos, as followers of Confucius would insist) we may cultivate the capacity and the will to speak the word that tells a story of and for and from within Nature—to say what the river says—and then stand by it. Then we can see the earth not as our adversary, nor as raw material, but as our only, most desirable, and most mysterious home, the only place where our humaneness may reach its full flower.

"But speak the truth, and all nature and all spirits help you with unexpected furtherance," says Emerson. "Whilst a man seeks good ends, he is strong by the whole strength of nature."[25] That's a story that can be finally and fully realized only through our own spiritual lives. It's a story of subsistence that will create a sustainable life, restore balance to the environment, and bring harmony to all our societies in a fashion so sound and fair that "justice will roll down like water" (Amos 5:24) for all creation. There we have, just as in Koyukon tradition, the natural and the spiritual inseparable. We share, then, a task that is both spiritual and pragmatic and a path, a way that leads toward sustainability. Both the task and the path call us to create ourselves, to cultivate ourselves so that we can grow into the best selves we can become. Our children's and our grandchildren's lives depend upon our response.

Conclusion

Creating a Sustainable Culture

"WE NEED TO TELL YOU up front that we are not here to serve this American culture that surrounds us, but to transform it—little by little —across the region we live in, and other places we can afford to reach or influence." That is the opening sentence of a proposal our little non-profit, the Northern Plains Sustainable Agriculture Society (NPSAS), sent to a foundation recently. I wrote that with my heart in my fingers and some fear in my heart. Most nonprofits and foundations are about service of some kind. Churchgoing Christian friends desire to transform the culture as well, to save it and all of us for heaven. Since I am not a churchgoing person and find visions of heaven and hell equally unattractive, I do not have much interest in salvation and am therefore free to ask some of the questions that Jesus raised, such as, Who is my neighbor? One has to take such a question personally, locally. I have to give my answer every day in my response to people around me, yes, but also to other creatures, the nearby lakes, streams, fields, forests, and their inhabitants. Whatever exists in my neighborhood is my neighbor—even that fool woodpecker I saw this morning, stabbing away frenetically at a tall, silver light pole, raising a metallic din that clobbered the morning stillness like a jackhammer. What did she expect to find? Didn't her momma teach her right? Regardless of her poor upbringing, she is my neighbor, although there is not much I can do about her beak, which must surely be blunted and benumbed by her fruitless, hammering search, except to chase her away, hopefully to something more rewarding, like a tree. But maybe she was a percussionist come back, and the resounding clang was music to her feathered ears.

At the moment my real work is as an administrator with NPSAS. We have more than four hundred members in twenty-some states and are out to transform the culture for the very reason that Prescott Bergh challenged us with in the introduction: "There's not much point in talking about sustainable agriculture if you don't have a sustainable culture to back it up." We do not expect to make huge strides with a task so daunting, even with willing and more powerful colleagues with whom we can collaborate. So we have been asking ourselves some questions. For example, How can a nonprofit that operates according to the culture's values ever hope to change that culture? Our conclusion is that it cannot. If we exhibit our culture's primary values in our own behavior, we can never have a transforming effect on the culture.

Considering this has led to another question: since growth and speed are two of our culture's greatest and least sustainable values, we have to ask ourselves what it means for us to "grow." We do not believe, as our culture seems to, that it means growing in size, adding staff, expanding offices, and increasing budgets. We no longer believe in the values that a culture of growth and speed represents, and we do not want to participate in it by getting much bigger ourselves. Mere growth in size does not represent a model for a sustainable future; growth in insight, understanding, and influence might. We are not saying that we should never add another staff person or seek more funds. We are saying that our growth should be like our best farming—sustainable. Sustainable growth that emerges naturally, organically, in due season can teach us not only how to farm but also how to run a nonprofit and how to grow a sustainable culture. One of the core values of a sustainable agriculture is neighborliness. Who is our neighbor, indeed? If NPSAS can learn to operate in sustainable, neighborly ways, we will have a healthy nonprofit. We may not be rich or famous or have thousands of members, but we will continue to meet one of the great benchmarks for integrity: the story we tell the world and our behavior will match.

Our first task in transforming the culture is to transform ourselves, growing in our understanding of how the world really works at levels beneath the political, corporate, and public relations gloss that has been laid upon it. That is only a veneer beneath which we can see the losses that scar an unsustainable culture: loss of topsoil, loss of fertility in the soil that is left, loss of groundwater, loss of drinking water, loss of pollinators, loss of plant and animal diversity, loss of nutritional value in the plants and animals we still have, loss of small farms, loss of small towns, loss of our sense of community, a discernible decline in compassion and empowerment. I could go on here, but you know these things in your own experience. Getting bigger and bigger would not be a sign of our success but would indicate that we have succumbed to our culture's more destructive fantasies and lost the discernment necessary to change it.

The healthy story NPSAS and our nonprofit allies now have to tell is about sustainable and organic agriculture. If there still is an America in 2050, it will exist, partly at least, because the culture has finally adopted a new story that there is a healthier way to farm and that we can create healthier food systems that are environmentally sound, socially just, and economically viable. Those agrarian principles we practice will also work to create a healthy society and a healthy economy for all of us, not just a few.

NPSAS probably can't "accomplish" that, but that does not mean that it is a fool's errand to try. What we can do is make our contribution to the creation of a sustainable culture, and we can collaborate with others who have the same goal. Our strategy is a bit different from that of many others who are out for cultural change. Our task is not to get big or create huge, "successful" projects whose every aspect is measurable but to intentionally multiply smallnesses, to generate enough sustainable smallnesses that they begin to connect with others in relationships that are sustainable. In fact we no longer have the luxury of working on "projects." Projects are but fragments, and we have Humpty Dumpty on our hands. The fragments won't restore Mr. Dumpty. We have to develop more holistic strategies to get ahead of the forces that are converging upon us: global warming, oil depletion, the loss of local food systems, water depletion, increasing violence around the globe, increasing hunger . . .

Our agribusiness giants brag about feeding the world. Their advertisements feature that success in brilliant color and describe it using the finest PR blurbs that money can buy. That colorful, full-page ad is deceit of the most dangerous kind: self-deceit deliberately designed to deceive the rest of us. We know that we do not even feed our own citizens very well, let alone others in the world. Estimates of the number of children without food security in the United States range from 12.5 million to 13.2 million to who knows how many who have not been identified but will go hungry this week. That's millions, yes. There is a lot of talk these days, at last, about local food, and like many others I cherish that as another sign of hope. And I hear talk about farmers finally getting paid a fair price for their produce, and I can only applaud that. But I do not hear much talk about how to match local food with local hunger. If a local food system feeds only high-end restaurants and middle- and upper-class families, how is it superior to the system we have, which leaves hungry children out of its equations? I do hear that when there are leftovers, some farmers' markets take their food to charitable organizations that do serve hungry citizens. That is fine but inadequate; we need to make the hungry a priority, the first receivers of local food. The rest of us can do with their leftovers.

It is not just the increasing number of hungry children we should think about but the increasing amount of food lacking in nutritional value, with the evidence mounting that the vegetables we used to rely upon to nourish our bodies are less effective than they were fifty years ago, and less pleasing to our palates. So now I have to eat two or three times the amount of brussels sprouts and broccoli to stay healthy? What kind of nutritional world is that?

Think how different our rural countryside would appear today if over the last half century we had been multiplying smallnesses rather than stimulating bigness. There would be flourishing small farms, prosperous small communities, healthier food on the table, and a means by which to feed ourselves and the world.

As partners in a concerted effort, NPSAS can yet increase farmers', teachers', social workers', civil servants', custodians', waitresses', bankers', and businesspeople's understanding of what it might take to create a sustainable society, and help it grow toward that sustainability. That NPSAS is small is no sign that our understanding or our vision is therefore small or that we cannot tackle big tasks and contribute to their accomplishment, working with others in a common cause.

We believe that our family farm members, and those colleagues in other sustainable and organic agricultural nonprofits, are a saving remnant akin to Isaiah's, buried under our culture's preoccupations with growth and power, speed and quick fixes. We do not need Madison Avenue's blaring horns or a full symphony to announce our message of sustainability; we can organize quiet conversations around small tables in our libraries in small towns and urban centers where everyone is welcome and the coffee is ready. We are here for the long haul, not for the quick fix; we aim for the incremental shifts in worldview that may not be immediately visible or make a noticeable splash but that testimony from participants in our programs will document for us. We did sixteen such meetings in 2007, including a series of four small conversations in the libraries in Fosston, Ada, and Pelican Rapids, Minnesota. One participant's evaluation read in part, "We need to continue and spread this discussion. It deals with our *whole being* and relations to all in the community and the world. This is a pivotal area in the life of people and creation—we need as much support, discussion and direction as possible to see how to go." Another participant wrote, "I'd like to have an ongoing dialogue with this group—maybe one time per month. I'd also like to have this same opportunity offered in more communities in northwest Minnesota. I'd like to have the next sessions be specifically designed to teach others to replicate." A third reported, "The program has rejuvenated my juices on this topic. I learned a few things along the way." But perhaps the best evaluation we'll ever get was this one from an organic sheep grower: "I came looking for 'how to' info and left with more questions than I came with. I'm pleased. I got what I needed instead of what I wanted."

Diverse small groups in intense discussions on related agrarian themes benefit from a kind of cultural osmosis. The first thing we want to do after we've heard something stirring, seen something beautiful, or read something important is tell somebody about it. The word spreads; the influence of those conversations seeps through the membranes and spills across the walls that often separate us. I believe it spreads best from small groups like these, small nodes of conversation and influence in which participants' lives and thoughts are changed by engaged and frequently intense conversation, good reading, and expanding ideas and understandings. They happen best as they occur in the little library groups we met with in February and March 2007. We'd rather reach a hundred folks a dozen at a time with transforming conversations than reach a thousand who experience a brief, virtual high via television, and whose interest shifts immediately with the next commercial. The kind of smallnesses we have in mind will allow NPSAS to, bit by bit, story by story, as best we can in the small ways open to us, transform the culture.

By multiplying such smallnesses, we can change the stories our culture tells itself. Our stories are more important agents of change than information and argument. We have known about climate change and oil depletion for decades and have not changed our behavior. Argument further divides us and creates defensiveness and aggression. But with the story NPSAS is telling, it can become one of those five smooth stones that will bring down the swaggering transnational Goliaths our culture idolizes and increase the odds for our children's and grandchildren's survival.

We believe that enough of the smaller cultural transformations toward greater sustainability NPSAS has to offer will ultimately increase the number of viable small farms, increase the biodiversity of our plants and pollinators, and result in smaller acreages with a greater variety of crops and fewer acres devoted to monocultured commodities. Finally such small events, sharing the stories we can tell through NPSAS, will increase the control over our own destiny that individual farmers deserve but have largely lost to the pressures of farming for the farm bill or the dictates of a corporate contract.

These questions and understandings grow, in part at least, out of pondering our mission statement: "We are committed to the development of a more sustainable society through the promotion of ecologically sound, socially just, and economically viable food systems." That might seem to be task enough, but we have learned over the past quarter century that our real mission lies even deeper. Bergh's comment still lingers. So the mission behind the mission has become to create a sustainable culture, not just a sustainable agriculture or sustainable communities or sustainable economies. We are all in this together, and our urban companions and rural families have to reconnect with one another and with the land to reflect on our environment, our present circumstances, our history, our policies, and our actions so that we have the means to experience the earth and our role on it from a new perspective.

The difference between a culture of exploitation and a sustainable culture lies more in worldview than in practice. Practice always follows on worldview; change the worldview and a change in practice is soon to follow. Our real worldview is most often expressed in the story we tell ourselves about who we are and what we are about in the world. Our culture needs a new story, one that will lead to that new worldview, and NPSAS has one—a story of sustainability that our farm friends already know by heart.

In light of that, we renewed our participation in advocacy for sustainable and organic agriculture, monitored legislation and policy as the language of the 2007 farm bill emerged, and increased our communication among staff and board and constituents about the new policy as it unfolded. We published *Handbook and Field Guide to the 2007 Farm Bill* and distributed it to our members and friends, alerting them to changes that would better serve agriculture and the whole culture. We have resumed our collaborations with the Midwest Sustainable Agriculture Working Group, the Sustainable Agriculture Coalition, and the Center for Rural Affairs.

We provide education through our summer symposium that puts participants, we hope some of them urban, in direct contact with the land and with those who are creating sustainable practices that work to give us all a sustainable future. We make field day visits to farms to educate us all, and we will also support the field days of other sustainable agriculture associations. We also educate our members through our winter conference, with its outstanding keynote speakers and multiple small workshops led by knowledgeable experts in livestock, grains, vegetables, marketing, seed saving, weed control, certification requirements, federal policy, and junkyard forages for our kids' artwork. With our quarterly newsletter, *The Germinator*, we keep our constituents in touch with the news about one another, new events, discoveries and ideas in agriculture, and emerging policy issues.

Our Farm Breeder Club (FBC), having grown out a new drought- and fusarium-tolerant strain of wheat (FBC-Dylan), will register the strain in the public domain, an activity that seems to put our land grant university and transnational corporate teeth on edge. It apparently makes them wonder what trade-related intellectual patents are for if not profit. We had 1,650 bushels of seed available in 2007, and we had requests for more than 2,500 from farmers who were willing to help us grow out more. We are continuing our experiments with emmer, will grow out new seed enough for field trials and taste tests, and will eventually register it in the public domain and win state certification. Our plan then is to move to winter triticale, rye, buckwheat, and oats—seeds that are now considered specialty grains in the United States but are seen as among the world's great grains elsewhere. Putting competitive grains in the public domain gives every farmer access to seed, offers independent farmers a way to get around Monsanto, and increases plant diversity throughout the northern plains. A 10 percent royalty on the net from that seed will be split evenly between FBC and NPSAS—a small step toward sustainability for both.

My Neighbor's Acre is an NPSAS fund built on our desire to practice being good neighbors by assisting those who are suffering economically because of drought, flood, fire, tornado, or other calamity. Members who wish to can contribute to the fund, which people in difficulty can tap either by nomination or by their own request. Aside from the nominator, no one needs to know the name of the recipient, who can receive up to $2,500 in emergency cash without filling out an application or signing any papers. There is no requirement that recipients pay the money back. If they get back on their feet and so desire, they can contribute to My Neighbor's Acre and help someone else. Last year the fund sent checks to a family whose house burned down, a rancher besieged by drought, and another rancher who through a series of bad breaks could not pay the taxes on his farm.

We will educate and enlist new urban friends for sustainability with a series of programs in our urban centers in the Great Plains: "Why Should I Care? Agriculture for Urbanites." Alas, one hundred letters from urban allies in Fargo and Bismarck have greater impact than three hundred letters from farmers scattered across three states— but the four hundred letters combined have still greater effect. We cannot change the policy or the culture alone; we need each other, knowledgeable farmers and all the urban friends we can recruit and educate. Thus we will also extend our small discussion groups to other small town and urban libraries in other states. We call these gatherings "Up for Discussion: Conversations for the Fearless and the Not-So-Fearless."

We are only one of a number of nonprofits seeking the same ends. By multiplying such smallnesses, we gain another means to get a new story out to small town and urban folks, many of whom have been away from the farm so long they may no longer have a clue about how healthy food is grown or what concepts, ideas, and policies might support a sustainable future for all our citizens. We are making some friends and reminding folks who may have forgotten that without small farms we lose our small towns and add another blight to our rural landscape.

We believe all these activities have the same focus: creating a sustainable culture and creating an environmentally sound, socially just, and economically sustainable agriculture and food system, beginning with educating and transforming ourselves. We believe these activities give us a coherent approach across a variety of cultural fronts to create social justice, increase public health, create or maintain local food systems that will help us survive the end of cheap oil, ease up on our water pollution and depletion, and restore our small communities' economic health at the same time. These are not special "projects," created to entice funds out of some foundation. This is who we are, a diverse group of plain folks with a common mission that unites us, and this is what our efforts have become—a way of life. This little institution has been engaged in this worthy task for almost three decades now.

This effort has not become our way of life by accident but because we have been thinking about it, deliberately trying to create a coherent, holistic, comprehensive set of activities that are anchored in the daily pragmatics of farm life and yet address the larger policy issues that ultimately affect us as producers and consumers. We are trying to make some contribution—modest though it is—toward the transformation of the culture. We believe all these activities are inextricably interrelated. The sum of these activities is a new, alternative story we are trying to live out ourselves and share with the larger culture. It is a story of cooperation instead of competition, of connections rather than fragments, of respect instead of condescension, of openness instead of secrecy, of good health and advantage for all instead of extravagant profit for a few. It is a story that creates a worldview that can lead us toward a sustainable society. We have children. We have grandchildren. Some of us have great-grandchildren. We need to share this new story with each other and with them. If we do not, they will have no future at all.

At NPSAS we pursue this way of life not because we think we will prevail; we can never be assured of that. We live this effort because, successful or not, it seems like the right thing to do. So we, all of us together, do it.

For Those Who Love Nature

A Benediction Antiphon

For all Nature's gifts,
All praise:

For our sun, and all other suns.
For rain as it comes to us, and
For drought and flood.
    For they are all natural

For all animals, fierce or friendly.
For birds and butterflies and reptiles and
For all species, everywhere and every kind.
    For we are all related

For our human friends and loves:
For those we fear and those we hate.
For all saints and all murderers and
For all those who are neither.
    For we are all related

For all languages,
    For they bring the world to us
For all the arts,
    For they restore our vision

For all knowledge wise enough to confess ignorance,
For that keeps us real

For human degradations when they happen to us,
For they are part of a cycle
For our spirit's exaltations whenever they occur,
For they are part of a cycle
For disgrace and renewal,
For they, too, are part of a cycle
For depression and the lifting of burdens,
For they are part of a cycle

For pain and joy.
For rage and peace.
For growth and age.
For fear and courage.
For they are all natural, and remind us we are alive

For canyons, mountains, rivers, plains, plants, and rocks,
For they are all natural
For urban alleys, greasy Dumpsters, and concrete,
For they are all natural
For rats, vermin, vomit, plastic,
For they, also, are natural

For every risk that makes us grow,
For cancers, AIDS, and all other diseases,
For they are natural
For every chance we have to heal another,
Ourselves, or the Earth, and
For everything that helps us find our place
And leads us home
For all these too are natural

For life,
For death . . .

For Nature,
    All One,
    All accepting
        For all of it

All of it . . .

All praise . . .

                —In collaboration with C. J. Taylor

Acknowledgments

Many friends have contributed to my thinking about these matters, some for a long time. I begin with those to whom this book is dedicated. Bob Arnold, first director of the Alaska Public Broadcasting Commission, first consultant to Alaska's village schools about bilingual education, former director of the Center for Equal Opportunity in Schooling, was also for a time responsible for enabling the efforts of the Bureau of Land Management to meet its obligations to implement the transfer of land title to native people under the Alaska Native Claims Settlement Act. His views and ideas lurk in these pages as surely as my own. Dorik Mechau, cochair of the Island Institute in Sitka, Alaska, has also been friend, inspiration, and example for more than thirty years. He, Bob Arnold, and I were fortunate to meet together, work together, and talk together often. Dorik has been a partner in many adventures, ranging from fly-in fishing to fencing with educational institutions and state and federal governments. His views of the matters under consideration here have shaped my own. Ted Chamberlin, professor of comparative literature at the University of Toronto, has worked with First Nations people on land claims and land use issues for many years. He has also worked hard to understand the power of language, especially oral traditions, not only in Canada but in Africa, Australia, the Caribbean, Ireland, and Scotland. His experience, and our conversations, often exuberant, have always pushed me to think harder about things. The thoughtful approach of these men to the world, and to their own lives, certainly improved my understanding of the world, and of my own life.

For almost as long, poet Gary Snyder's views, expressed in his writing and in conversations over the years in places as diverse as California, Alaska, Montana, and Texas, have helped clarify my thinking. He read parts of this manuscript and offered helpful comments on it. Richard Nelson, a writer and anthropologist with long, insightful experience among Arctic and sub-Arctic indigenous peoples, has influenced me with his work and his conversations, readily sharing his views of subsistence and native life generally. Years ago, linguists Ron and Suzi Scollon shared with me their insights into indigenous languages and customs, enabling my own observations. I benefited greatly in those early years from conversations with anthropologist Ann Fienup-Riordan as she began her ventures into Yup'ik country, learned the language, and began to write. Her work led to more of her own investigations and collaborations in Yup'ik country and additional research by others. I was also privileged to participate in lengthy conversations with Michael Krauss, longtime director of the Alaska Native Language Center, and his colleague Irene Reed, known in Yup'ik country as "the blond Yup'ik" for her skill in the language. Elsie Mather, Eliza Jones, Rachel Craig, Martha Demientieff, and other native linguists offered insight and friendship. I am grateful as well for the observations and insights of Richard and Nora Dauenhauer over many years working on bilingual and Alaska native literary issues. Carolyn Servid's friendship and work with the Island Institute has informed my own views and enabled many others to think harder about sustainability through the public programs the institute has sponsored. Sitka fiction writer and poet John Straley and his wife Jan, a cetologist, have offered their rich friendship and ideas through the years.

Biologist and writer Gary Nabhan, who has worked on these issues directly for years, read the entire manuscript and offered his insights and suggestions. Steven Epstein, professor of history at the University of Kansas, read a full draft, suggested changes, and offered helpful comments. His wife, Jean Epstein, joined us in many of those conversations. Carol Wilson, longtime director of the Colorado Partnership for Educational Renewal, has talked with me often about educational and social issues and provided her quiet guidance and suggestions. International consultant on issues of civility and public life David Chrislip read an early version of this work and is another profound influence. David Armstrong, a thoughtful biologist at the University of Colorado, offered a meticulous and telling critique of many of these ideas as they began to surface in an earlier paper and caused me to temper some of my notions about the natural world and humans' role in it.

I have talked over these matters at length with Episcopal priest and social and environmental activist Benjamin Webb. His encouragement enabled me to persist with this project and certainly broadened my views of both sustainability and faith. Mary Evelyn Tucker, cochair of the Forum on Religion and Ecology and now teaching at the Yale School of Forestry and Environmental Studies and at Yale Divinity School, has afforded me many opportunities to dig into Confucian works and shared her insights into Confucianism, other world religions, and ecology. Those insights, too, have helped shape my thinking. I continue to be amazed at how current and useful the works of Confucius, as well those of his followers at every period, and of the ancient Chinese poets from the earliest records, continue to be. They help balance my bent toward the pre-Socratics and other Greeks and Romans who laid the foundations of Western traditions and whom I still find wonderfully useful in thinking about our own time.

Barbara Wendland, publisher of the courageous church newsletter *Connections*, once again exercised her editorial wisdom on an early draft, making it easier to read and clarifying ideas. Avid reader and champion of social justice C. J. Taylor used her careful red pen to advantage as well, questioning things, catching mistakes I'd missed, and making helpful suggestions about structure.

On the homefront, Prescott Bergh, a Wisconsin farmer and entrepreneur, triggered much of this with his comment about sustainable cultures. Minnesota Department of Natural Resources biologist Larry Gates offered advice and encouragement. His broad overview of the connections among things biological, social, and political is an inspiration. Dean Harrington, president of the First National Bank of Plainview and a colleague in the Rural America Arts Partnership, read the manuscript with his careful eye and asked the right questions in his gentle manner. Jeff Gorfine and Richard Broeker, the former chair of the board of directors and the executive director of southeast Minnesota's Experiment in Rural Cooperation, respectively, have been readers, friends, and supporters whose advice has always been helpful and whose passion for justice never fades. I deeply regret that Dick will not know how his influence on this work came out, for he left his fully engaged life much too early, in 2005. Peggy Thomas, a farmers' market producer, read the manuscript and offered her comments. Ralph Lentz, a cattleman whose own sustainable land use practices have upset numerous experts and changed their views, has also, often, to my great pleasure and enlightenment, changed mine—though I'd hate to admit that to Ralph, and I continue to resist every foot of contested ground.

Despite the importance of the persons mentioned above, the roots of the thinking here go back to the members of eighth grade and senior high school classes I taught in Naknek, Alaska. They were my students in language arts, American history, world history, and Spanish. Eighth grader Archie Gottschalk perhaps triggered my interest but was instantly joined by other students, including his sister Glenda, Glen McCormick, Adelheid Herrmann, Norman Anderson, Glenda Wilson, and Raymond and Starlett Patterson. High school students Juliana Ansaknok, George Gottschalk, Ramona Matson, Johnny Holstrom, Sarah Anderson, Bruce Kihle, Christine Nekeferoff, Raymond Nekeferoff, Ron Monsen, Larry Ring, Mary Clark, Barbara Monsen, Ellen Aspelund, Gerry Herrmann, Paul Carlson, and Bill Tolbert all taught me so much about Bristol Bay that I am still trying to learn from their comments—and their silences.

Alaska Methodist University in Anchorage had two native students in 1968, but the next year there were more than twenty-five, and because I was on the staff, I was lucky to know all of them. Some were Yup'ik or Inupiat; some were Athapaskan; some were Tlingit. Others were Aleut, Haida, or Tsimshian. There were far too many to remember all of them, but among those who taught me were Augusta Sykes, Henry Allen, Grace Antoghame, Sylvester Ayak, Edgar Blatchford, Sandra Borbridge, Barbara Brady, Esther Garber, Sister Goodwin, Andrew Hope, Barbara Jacko, Laura Jacko, Carolyn Kalkins, Albert Kookesh, Hannah Loon, Sandra Merculief, Clark Millet, Alexie Morris, Mary Nanuwak, Henry Oyoumick, David Sam, Edna Ungudruk, and Ginny Walker. Others are now just first names: Helen and Pooch, and Arch of course was there for a time. I have been out of touch with most of these folks for two decades, yet as I call up each name, images and events rise to accompany them in my memory. I was honored to know them years ago and am honored again by their presence in my thoughts now.

I am especially grateful to Stephen M. Wrinn, director of the University Press of Kentucky, who first suggested that there might be a book like this, and his assistant Anne Dean Watkins. As usual, they have been encouraging and generous throughout. David Cobb, Hap Houlihan, and Mack McCormick have worn their customary patience and helpfulness with ease. Once again copyeditor Anna Laura Bennett has exercised her professional skill and extended her generous spirit and grace at every turn. This is a much better book because of her fine hand and eye for detail.

Finally, my wife, Lauren Pelon, who always listens patiently and adds her own thinking to mine, has supported this work throughout the long process. To say that I'm grateful to all these folks is akin to criminal understatement. They have done their best to keep me straightened out; whatever remains bent, confused, or wrong in this is, alas, all my own doing.

"Everything is collage, even genetics," says Michael Ondaatje. "There is the hidden presence of others in us, even those we have known briefly. We contain them for the rest of our lives, at every border we cross."[1]

Notes

Introduction

1. For a discussion of the Neo-Confucians, see *A Source Book in Chinese Philosophy*, comp. and trans. Wing-tsit Chan (Princeton, N.J.: Princeton University Press, 1963), 460–691.

2. Mary Evelyn Tucker, "The Philosophy of *Ch'i* as an Ecological Cosmology," in *Confucianism and Ecology: The Interrelation of Heaven, Earth, and Humans*, ed. Mary Evelyn Tucker and John Berthrong (Cambridge, Mass.: Harvard University Center for the Study of World Religions, 1998), 189.

3. Gary Holthaus, *From the Farm to the Table: What All Americans Need to Know about Agriculture* (Lexington: University Press of Kentucky, 2006), 29–117.

4. Confucius, *The Great Learning*, in *Source Book*, 84–94. Also see Confucius, *The Great Digest*, in *The Great Digest, The Unwobbling Pivot, The Analects*, trans. Ezra Pound (New York: New Directions, 1951), 27–91.

5. Confucius, *The Analects*, trans. David Hinton (Washington, D.C.: Counterpoint, 1998), xiii, 3, 139–40; Epictetus, *The Art of Living: The Classic Manual on Virtue, Happiness, and Effectiveness*, trans. Sharon Lebell (San Francisco: Harper, 1995), 51.

6. For earlier uses of the logos, see G. S. Kirk, J. E. Raven, and M. Schofield, *The Presocratic Philosophers: A Critical History with Selected Texts*, 2nd ed. (New York: Cambridge University Press, 1988), 187–212.

7. Richard Hugo, "Distances," in *Making Certain It Goes On: The Collected Poems of Richard Hugo* (New York: Norton, 1984), 434.

8. Peter Schwartz and Doug Randall, *An Abrupt Climate Change Scenario and Its Implications for United States National Security*, report prepared by the Global Business Network for the Department of Defense,

October 2003, 22 (see also 2, 3, 5, 16–22), available from http://www.gbn
.com/ArticleDisplayServlet.srv?aid=26231.

Music and Story

1. Wes Jackson, *Becoming Native to This Place* (Lexington: University Press of Kentucky, 1994), 26.

2. Ibid., 8.

3. Margaret Bourke-White, "Dust Changes America," *Nation*, May 22, 1935, 597–98.

4. Jackson, *Becoming Native*, 8.

5. Ivan Illich, *Shadow Work* (Boston, Mass.: Boyars, 1981), 9, 55–74.

6. On Fetterman, see U.S. Department of the Army, *American Military History, 1607–1953* (Washington, D.C., 1956), 282.

7. *Letter of Secretary of War*, 39th Cong., 2nd sess., Senate Executive Document 15 (Washington, D.C., 1866–1867), 4. For an early account of American injustices to Indians that includes Sherman's telegram, see George W. Mannypenny, *Our Indian Wards* (Cincinnati: Clarke, 1880), 195.

8. *Agayuliyararput: Kegginaqut, Kangiit-llu = Our Way of Making Prayer: Yup'ik Masks and the Stories They Tell*, trans. Marie Meade, ed. Ann Fienup-Riordan (Seattle: University of Washington Press, 1996), 11–25.

9. T. S. Eliot, "Choruses from the Rock I," in *Selected Poems* (New York: Harcourt Brace Jovanovich, 1964), 107.

10. Theo Colborn, Dianne Dumanoski, and John Peterson Myers, *Our Stolen Future: Are We Threatening Our Fertility, Intelligence, and Survival? A Scientific Detective Story* (New York: Plume, 1997), 138–39.

11. Miguel A. Altieri, "Enhancing the Productivity of Latin American Traditional Peasant Farming Systems through an Agroecological Approach," *Agroecology in Action*, http://www.agroeco.org/fatalharvest/articles/enhancing_prod_la_peasants.html.

12. Ibid. For a perspective that includes Latin America and far beyond, see Miguel A. Altieri, Peter Rosset, and Lori Ann Thrupp, "The Potential of Agroecology to Combat Hunger in the Developing World," *Agroecology in Action*, http://www.agroeco.org/doc/potential_of_agroecology.html.

Habitat for a Sustainable Culture

1. Marcus Tullius Cicero, *Selected Works*, trans. Michael Grant (New York: Penguin Books, 1971), 171, 168.

2. Lao Tzu, *Hua Hu Ching: The Unknown Teachings of Lao Tzu*, trans. Brian Walker (New York: HarperCollins, 1995), 96.

3. Ralph Waldo Emerson, "Nature [1836]," in *Selected Essays*, ed. Larzer Ziff (New York: Penguin Books, 1982), 79.

4. Thomas Carlyle, *A Carlyle Reader*, ed. G. B. Tennyson (Acton, Mass.: Copley, 2000), 24.

5. Jackson, *Becoming Native*, 24.

6. William Stafford, "Ask Me," in *The Way It Is: New and Selected Poems* (St. Paul, Minn.: Graywolf Press, 1999), 56.

7. Thomas Berry, *The Dream of the Earth* (San Francisco: Sierra Club Books, 1990), 35.

8. Stafford, "Where We Are," in *Way It Is*, 34.

Functional Cultures and Structural Cultures

1. Schwartz and Randall, *Abrupt Climate Change*.

Exploring Subsistence

1. Ann Fienup-Riordan, *Eskimo Essays: Yup'ik Lives and How We See Them* (New Brunswick, N.J.: Rutgers University Press, 1990), 47–48.

2. Ibid., 45–46.

3. Dietrich Bonhoeffer, *Letters and Papers from Prison*, rev. ed., ed. Eberhard Bethge (New York: Macmillan, 1967), 13.

4. Taisen Nobuhara, *The Brilliant Life of Munetada Kurozumi: A Philosopher and Worshipper of the Sun*, trans. Tsukasa Sakai and Kazuko Sasage (Tokyo: PMC, 1980), 14.

5. See Richard K. Nelson, *Make Prayers to the Raven: A Koyukon View of the Northern Forest* (Chicago: University of Chicago Press, 1983), and Fienup-Riordan, *Eskimo Essays*.

6. Thomas J. DiLorenzo, "Sherman's Final Solution: How Lincoln's Army 'Liberated' the Indians," 2003, http://www.scvcamp469-nbf.com/sherman.htm.

Education for Subsistence

1. For more stories on Naknek and the area, see Gary Holthaus, "Teaching Eskimo Culture to Eskimo Students: A Special Program for Secondary Schools in Bristol Bay" (proposal to Alaska Department of Education, 1968).

2. For the best summary of Sherrard's views, see Philip Sherrard, *The Eclipse of Man and Nature: An Enquiry into the Origins and Consequences*

of Modern Science (West Stockbridge, Mass.: Lindisfarne Press, 1987), 15–30. Also see John Garvey, "An Alienating Culture: No Sense of the Sacred," *Commonweal*, November 18, 1994.

3. Emerson, "Nature [1836]," 75.

4. Gary Snyder, *Axe Handles* (San Francisco: North Point Press, 1983), 59–114.

5. Kathleen Dean Moore, *The Pine Island Paradox* (Minneapolis: Milkweed Editions, 2004), 158.

Education for Sustainability

1. Eliza Jones (lecture, Sitka Summer Symposium, Sitka, Alaska, 1993).

2. Plato, *Meno*, in *Five Dialogues*, trans. G. M. A. Grube (Indianapolis: Hackett, 1981), 59–88.

3. Confucius, *Great Learning*, 86–87; Confucius, *Analects*, 40.

4. Lucius Annaeus Seneca, *Letters from a Stoic*, trans. Robin Campbell (New York: Penguin Books, 1969), 151–61.

5. John I. Goodlad, Roger Soder, and Kenneth A. Sirotnik, preface to *The Moral Dimensions of Teaching*, ed. John I. Goodlad, Roger Soder, and Kenneth A. Sirotnik (San Francisco: Jossey-Bass, 1990), xii–xiv; Kenneth A. Sirotnik, "Society, Schooling, Teaching and Preparing to Teach," in ibid., 308–14.

6. Plato, *The Republic*, trans. H. D. P. Lee (New York: Penguin Books, 1988), 229, 228–36.

7. Sam Keen, *To a Dancing God* (New York: Harper and Row, 1970), 41.

Imagining Sustainability

1. Mary Oliver, "Spring," in *New and Selected Poems* (Boston: Beacon Press, 1992), 70.

2. Henry George, "The Crime of Poverty" (address delivered in the Opera House, Burlington, Iowa, April 1, 1885), http://www.cooperativeindividualism.org/george-henry_crime-of-poverty.html.

3. Gary Snyder, "For the Children," in *Turtle Island* (New York: New Directions, 1974), 86.

4. World Commission on Environment and Development, *Our Common Future* (Oxford: Oxford University Press, 1987).

Defining Sustainability

1. J. Hector St. John de Crèvecoeur, *Letters from an American Farmer and Sketches of Eighteenth-Century America*, ed. Albert E. Stone (New York: Penguin Books, 1981), 56.

2. Samuel Sewall, *The Selling of Joseph: A Memorial* (Digital Commons, University of Nebraska–Lincoln, 2007), http://digitalcommons.unl .edu/etas/26/.

3. Albert Schweitzer, *Out of My Life and Thought: An Autobiography*, trans. C. T. Campion (New York: New American Library, 1953), 124–27.

4. Martin Buber, *I and Thou*, trans. Ronald Gregor Smith, 2nd ed. (New York: Scribner, 1958), 3–12.

Stories for Sustainability

1. Phyllis Morrow and William Schneider, introduction to *When Our Words Return: Writing, Hearing, and Remembering Oral Traditions of Alaska and the Yukon*, ed. Phyllis Morrow and William Schneider (Logan: Utah State University Press, 1995), 1.

2. Ibid., 2.

3. Elias Canetti, *The Conscience of Words*, trans. Joachim Neugroschel (New York: Farrar Straus Giroux, 1984), 6–7, 238–39.

4. William Wordsworth, preface to *Lyrical Ballads*, in *The Complete Poetical Works of William Wordsworth*, ed. Andrew J. George (Boston: Houghton Mifflin, 1932), 791. Wordsworth altered this preface each time the volume was reprinted, from 1800 to 1845. This version of his famous dictum is from the preface to the 1800 edition.

5. Tina Rosenberg, *The Haunted Land: Facing Europe's Ghosts after Communism* (New York: Vintage Books, 1996), xxiii–xxiv.

6. T. S. Eliot, "East Coker," in *Four Quartets* (New York: Harcourt Brace Jovanovich, 1971), 31.

7. Oscar Handlin, *Truth in History* (Cambridge, Mass.: Belknap Press, 1979), 405. For his complete argument, see 403–15.

8. Livy quoted in R. H. Barrow, *The Romans* (New York: Penguin Books, 1979), 86; Cicero, *Selected Works*, trans. Michael Grant (New York: Penguin Books, 1971), 192, 177–78, 184, 179.

9. Robert D. Putnam, *Bowling Alone: The Collapse and Renewal of American Community* (New York: Simon and Schuster, 2000).

10. Lowry quoted in Ted Bernard and Jora Young, *The Ecology of Hope: Communities Collaborate for Sustainability* (Gabriola Island, B.C.: New Society, 1997), 143.

11. Bernard and Young, *Ecology of Hope*, 129–33.

12. Ibid., 139.

13. Ibid.

14. Ibid., 144.

15. Ibid., 146.

16. Ibid., 147.

17. Nelson, *Make Prayers*, 240–41.

18. Hugo, "Distances," 434.

The Power and Pragmatism of Language

1. Confucius, *Analects*, 139–40.

2. Ibid., 140.

3. *Herakleitos and Diogenes*, trans. Guy Davenport (Bolinas, Calif.: Grey Fox Press, 1979), 11; Charles H. Kahn, *The Art and Thought of Heraclitus: An Edition of the Fragments with Translation and Commentary* (Cambridge: Cambridge University Press, 1993), 43.

4. Seyyed Hossein Nasr, *Religion and the Order of Nature* (New York: Oxford University Press, 1996), 92–93; *Herakleitos and Diogenes*, 22.

5. André Padoux, introduction to *Vāc: The Concept of the Word in Selected Hindu Tantras*, trans. Jacques Gontier (Albany: State University of New York Press, 1990), x, xiv, xi.

6. Nasr, *Religion*, 37.

7. *The Upanishads*, trans. by Juan Mascaró (New York: Penguin Books, 1979), 127; Padoux, introduction to *Vāc*, xi; Fienup-Riordan, *Eskimo Essays*, 211.

8. Confucius, *Great Digest*, 33, 20.

9. Lao Tzu, *Tao Te Ching*, trans. Gia-fu Feng and Jane English (New York: Vintage, 1987), 3.

10. *The Confessions of Saint Augustine*, trans. Rex Warner (New York: New American Library, 1961), 17; James Hillman, *Healing Fiction* (Barrytown, N.Y.: Station Hill Press, 1983), esp. "What Does the Soul Want?" 85–129; Victor Frankl, *Man's Search for Meaning*, rev. ed. (New York:

Washington Square Press, 1985), 121. See also Victor Frankl, *The Unheard Cry for Meaning: Psychotherapy and Humanism* (New York: Washington Square Press, 1985), 19.

11. Bill McKibben, *The End of Nature* (New York: Random House, 2006).

12. Ronald Reagan quoted in Kevin Sullivan and Mary Jordan, *Washington Post*, June 10, 2004.

13. N. Scott Momaday, *House Made of Dawn* (New York: Harper and Row, 1989), 58, 59.

14. Ibid., 92–98; N. Scott Momaday, *The Way to Rainy Mountain* (Albuquerque: University of New Mexico Press, 2005), 33.

15. Momaday, *House Made of Dawn*, 209, 211, 2, 212.

16. Gregory Cajete, *Look to the Mountain: An Ecology of Indigenous Education* (Asheville, N.C.: Kivaki Press, 1994), 42; Lao Tzu, *Hua Hu Ching*, 106.

17. Simon Ortiz, "Song/Poetry, and Language—Perception and Expression," in *Speak to Me Words: Essays on Contemporary American Indian Poetry*, ed. Dean Rader and Janice Gould (Tucson: University of Arizona Press, 2003), 237.

18. Arthur Schopenhauer, *Essays and Aphorisms*, trans. R. J. Hollingdale (New York: Penguin Classics, 1976), 205.

Rectifying the Names

1. Evelyn Underhill, *The Spiritual Life* (New York: Harper, 1936), 11–12.

2. Robert Cummings Neville, *Boston Confucianism: Portable Tradition in the Late-Modern World* (Albany: State University of New York Press, 2000), 63, 61, 63.

3. Martin Buber, *The Way of Response*, ed. N. N. Glatzer (New York: Schocken Books, 1966), 64.

4. *Parmenides and Empedocles: The Fragments in Verse Translation*, trans. Stanley Lombardo (San Francisco: Grey Fox Press, 1982), 53, 54, 56, 61; *Herakleitos and Diogenes*, 42, 56, 31, 11.

5. Martin Buber, *To Hallow This Life: An Anthology*, ed. Jacob Trapp (Westport, Conn.: Greenwood Press, 1958), 90, xiii, 91.

6. Martin Rees, *Just Six Numbers: The Deep Forces That Shape the Universe* (New York: Basic Books, 2000), 51.

7. Bonhoeffer, *Letters and Papers*, 175–204.

8. Dietrich Bonhoeffer, *Selections from His Writings*, ed. Eileen Taylor (Springfield, Ill.: Templegate, 1992), 71.

9. Oliver, "Spring," 70.

10. Schweitzer, *Out of My Life*, 125.

A Spirituality for Our Time

1. Herman E. Daley, "The Lurking Inconsistency," *Conservation Biology* 13 (1999): 693–94.

2. Ralph Waldo Emerson, "Nature [1849]," in *Essays and Lectures* (New York: Library of America, 1983), 551–52.

3. Richard Poirier, *Robert Frost: The Work of Knowing* (New York: Oxford University Press, 1977), 278.

4. Buber, *Way of Response*, 63.

5. Kitaro Nishida, *An Inquiry into the Good*, trans. Masao Abe and Christopher Ives (New Haven, Conn.: Yale University Press, 1990), 3–10, 73–83, 57–59.

6. Robinson Jeffers, "Carmel Point," in *Selected Poems* (New York: Vintage, 1965), 102; Robinson Jeffers, "Their Beauty Has More Meaning," in ibid., 77; Robinson Jeffers, "The Inhumanist 5," in *The Double Axe and Other Poems Including Eleven Suppressed Poems* (New York: Liveright, 1977), 54.

7. Emerson, "Nature [1836]," 72–73.

8. Sara Teasdale, "There Will Come Soft Rains (War Time)," in *The Collected Poems of Sara Teasdale* (New York: Macmillan, 1938), 189.

9. *The Gary Snyder Reader: Prose, Poetry, and Translations, 1952–1998* (Washington, D.C.: Counterpoint, 1999), 260.

10. T. S. Eliot, "The Hollow Men," in *Selected Poems*, 80.

11. Mary Evelyn Tucker, introduction to *The Philosophy of Qi: The Record of Great Doubts* by Kaibara Ekken, trans. Mary Evelyn Tucker (New York: Columbia University Press, 2007), 58–60.

12. *Herakleitos and Diogenes*, 11.

13. Mary Evelyn Tucker, *Moral and Spiritual Cultivation in Japanese Neo-Confucianism: The Life and Thought of Kaibara Ekken* (Albany: State University of New York Press, 1989), 31–51, 136.

14. Ibid., 138.

15. Ibid., 139, 140.

16. Nelson, *Make Prayers*, 240.

17. T. S. Eliot, "Burnt Norton," in *Four Quartets*, 15–16.

18. Young-chan Ro, *The Korean Neo-Confucianism of Yi Yulgok* (Albany: State University of New York Press, 1989), 106–8.

19. Buber quoted in Maurice Friedman, *Martin Buber and the Eternal* (New York: Human Sciences Press, 1986), 149.

20. Nobuhara, *Munetada Kurozumi*, 49–50.

21. Johan Bojer, *The Great Hunger*, trans. Charles Archer and William John Alexander Worster (Project Gutenberg, 2006), http://www.gutenberg.org/etext/2943.

22. Thomas Merton, *The Sign of Jonas* (Garden City, N.Y.: Image Books, 1956), 267; Berry, *Dream of the Earth*, 195; Nishida, *Inquiry into the Good*, 158.

23. Koyukon elder quoted in Nelson, *Make Prayers*, 241.

24. *The Wisdom of Confucius*, ed. and trans. Lin Yutang (New York: Modern Library, 1994), 124.

25. Ralph Waldo Emerson, "An Address Delivered before the Senior Class in Divinity College, Cambridge, Sunday Evening, July 15, 1838," in *Essays and Lectures*, 77.

Acknowledgments

1. Michael Ondaatje, *Divisadero* (New York: Knopf, 2007), 16.

Bibliography

Agayuliyararput: Kegginaqut, Kangiit-llu = Our Way of Making Prayer: Yup'ik Masks and the Stories They Tell. Translated by Marie Meade. Edited by Ann Fienup-Riordan. Seattle: University of Washington Press, 1996.

Altieri, Miguel A. "Enhancing the Productivity of Latin American Traditional Peasant Farming Systems through an Agroecological Approach." *Agroecology in Action.* http://www.agroeco.org/fatalharvest/articles /enhancing_prod_la_peasants.html.

Altieri, Miguel A., Peter Rosset, and Lori Ann Thrupp. "The Potential of Agroecology to Combat Hunger in the Developing World." *Agroecology in Action.* http://www.agroeco.org/doc/potential_of_agroecology.html.

Augustine. *The Confessions of Saint Augustine.* Translated by Rex Warner. New York: New American Library, 1961.

Barrow, R. H. *The Romans.* New York: Penguin Books, 1979.

Bernard, Ted, and Jora Young. *The Ecology of Hope: Communities Collaborate for Sustainability.* Gabriola Island, B.C.: New Society, 1997.

Berry, Thomas. *The Dream of the Earth.* San Francisco: Sierra Club Books, 1990.

Bojer, Johan. *The Great Hunger.* Translated by Charles Archer and William John Alexander Worster. Project Gutenberg, 2006. http://www.guten berg.org/etext/2943.

Bonhoeffer, Dietrich. *Letters and Papers from Prison.* Rev. ed. Edited by Eberhard Bethge. New York: Macmillan, 1967.

———. *Selections from His Writings.* Edited by Eileen Taylor. Springfield, Ill.: Templegate, 1992.

Bourke-White, Margaret. "Dust Changes America." *Nation,* May 22, 1935, 597–98.

Buber, Martin. *I and Thou.* 2nd ed. Translated by Ronald Gregor Smith. New York: Scribner, 1958.

———. *To Hallow This Life: An Anthology.* Edited by Jacob Trapp. Westport, Conn.: Greenwood Press, 1958.

———. *The Way of Response.* Edited by N. N. Glatzer. New York: Schocken Books, 1966.

Cajete, Gregory. *Look to the Mountain: An Ecology of Indigenous Education.* Asheville, N.C.: Kivaki Press, 1994.

Canetti, Elias. *The Conscience of Words.* Translated by Joachim Neugroschel. New York: Farrar Straus Giroux, 1984.

Carlyle, Thomas. *A Carlyle Reader.* Edited by G. B. Tennyson. Acton, Mass.: Copley, 2000.

Cicero, Marcus Tullius. *Selected Works.* Translated by Michael Grant. New York: Penguin Books, 1971.

Colborn, Theo, Dianne Dumanoski, and John Peterson Myers. *Our Stolen Future: Are We Threatening Our Fertility, Intelligence, and Survival? A Scientific Detective Story.* New York: Plume, 1997.

Confucius. *The Analects.* Translated by David Hinton. Washington, D.C.: Counterpoint, 1998.

———. *The Great Digest, The Unwobbling Pivot, The Analects.* Translated by Ezra Pound. New York: New Directions Books, 1951.

Daley, Herman E. "The Lurking Inconsistency." *Conservation Biology* 13 (1999): 693–94.

DiLorenzo, Thomas J. "Sherman's Final Solution: How Lincoln's Army 'Liberated' the Indians." 2003. http://www.scvcamp469-nbf.com/sherman .htm.

Ekken, Kaibara. *The Philosophy of Qi: The Record of Great Doubts.* Translated by Mary Evelyn Tucker. New York: Columbia University Press, 2007.

Eliot, T. S. *Four Quartets.* New York: Harcourt Brace Jovanovich, 1971.

———. *Selected Poems.* New York: Harcourt Brace Jovanovich, 1964.

Emerson, Ralph Waldo. *Essays and Lectures.* New York: Library of America, 1983.

———. *Selected Essays.* Edited by Larzer Ziff. New York: Penguin Books, 1982.

Epictetus. *The Art of Living: The Classic Manual on Virtue, Happiness, and Effectiveness.* Translated by Sharon Lebell. San Francisco: Harper, 1995.

Fienup-Riordan, Ann. *Eskimo Essays: Yup'ik Lives and How We See Them.* New Brunswick, N.J.: Rutgers University Press, 1990.

Frankl, Victor. *Man's Search for Meaning.* Rev. ed. New York: Washington Square Press, 1985.

———. *The Unheard Cry for Meaning: Psychotherapy and Humanism.* New York: Washington Square Press, 1985.

Friedman, Maurice. *Martin Buber and the Eternal.* New York: Human Sciences Press, 1986.

Goodlad, John I., Roger Soder, and Kenneth A. Sirotnik, eds. *The Moral Dimensions of Teaching.* San Francisco: Jossey-Bass, 1990.

Handlin, Oscar. *Truth in History.* Cambridge, Mass.: Belknap Press, 1979.

Herakleitos and Diogenes. Translated by Guy Davenport. Bolinas, Calif.: Grey Fox Press, 1979.

Hillman, James. *Healing Fiction.* Barrytown, N.Y.: Station Hill Press, 1983.

Holthaus, Gary. *From the Farm to the Table: What All Americans Need to Know about Agriculture.* Lexington: University Press of Kentucky, 2006.

Hugo, Richard. *Making Certain It Goes On: The Collected Poems of Richard Hugo.* New York: Norton, 1984.

Illich, Ivan. *Shadow Work.* Boston, Mass.: Boyars, 1981.

Jackson, Wes. *Becoming Native to This Place.* Lexington: University Press of Kentucky, 1994.

Jeffers, Robinson. *The Double Axe and Other Poems Including Eleven Suppressed Poems.* New York: Liveright, 1977.

———. *Selected Poems.* New York: Vintage, 1965.

Kahn, Charles H. *The Art and Thought of Heraclitus: An Edition of the Fragments with Translation and Commentary.* Cambridge: Cambridge University Press, 1993.

Keen, Sam. *To a Dancing God.* New York: Harper and Row, 1970.

Kirk, G. S., J. E. Raven, and M. Schofield. *The Presocratic Philosophers: A Critical History with Selected Texts.* 2nd ed. New York: Cambridge University Press, 1988.

Lao Tzu. *Hua Hu Ching: The Unknown Teachings of Lao Tzu.* Translated by Brian Walker. New York: HarperCollins, 1995.

———. *Tao Te Ching.* Translated by Gia-fu Feng and Jane English. New York: Vintage, 1987.

Lee, David. *News from Down to the Café.* Port Townsend: Copper Canyon Press, 1999.

Letter of Secretary of War. 39th Cong., 2nd sess. Senate Executive Document 15. Washington, D.C., 1866–1867.

McKibben, Bill. *The End of Nature.* New York: Random House, 2006.

Merchant, Carolyn. *The Death of Nature: Women, Ecology, and the Scientific Revolution.* San Francisco: Harper and Row, 1990.

Merton, Thomas. *The Sign of Jonas.* Garden City, N.Y.: Image Books, 1956.

Momaday, N. Scott. *House Made of Dawn.* New York: Harper and Row, 1989.

———. *The Way to Rainy Mountain.* Albuquerque: University of New Mexico Press, 2005.

Moore, Kathleen Dean. *The Pine Island Paradox.* Minneapolis: Milkweed Editions, 2004.

Morrow, Phyllis, and William Schneider, eds. *When Our Words Return: Writing, Hearing, and Remembering Oral Traditions of Alaska and the Yukon.* Logan: Utah State University Press, 1995.

Nasr, Seyyed Hossein. *Religion and the Order of Nature.* New York: Oxford University Press, 1996.

Nelson, Richard K. *Make Prayers to the Raven: A Koyukon View of the Northern Forest.* Chicago: University of Chicago Press, 1983.

Neville, Robert Cummings. *Boston Confucianism: Portable Tradition in the Late-Modern World.* Albany: State University of New York Press, 2000.

Nishida, Kitaro. *An Inquiry into the Good.* Translated by Masao Abe and Christopher Ives. New Haven, Conn.: Yale University Press, 1990.

Nobuhara, Taisen. *The Brilliant Life of Munetada Kurozumi: A Philosopher and Worshipper of the Sun.* Translated by Tsukasa Sakai and Kazuko Sasage. Tokyo: PMC, 1980.

Oliver, Mary. *New and Selected Poems.* Boston: Beacon Press, 1992.

Ondaatje, Michael. *Divisadero.* New York: Knopf, 2007.

Padoux, André. Introduction to *Vāc: The Concept of the Word in Selected Hindu Tantras.* Translated by Jacques Gontier. Albany: State University of New York Press, 1990.

Parmenides and Empedocles: The Fragments in Verse Translation. Translated by Stanley Lombardo. San Francisco: Grey Fox Press, 1982.

Plato. *Five Dialogues.* Translated by G. M. A. Grube. Indianapolis: Hackett, 1981.

———. *The Republic.* Translated by H. D. P. Lee. New York: Penguin Books, 1988.

Poirier, Richard. *Robert Frost: The Work of Knowing.* New York: Oxford University Press, 1977.

Putnam, Robert D. *Bowling Alone: The Collapse and Renewal of American Community*. New York: Simon and Schuster, 2000.

Rader, Dean, and Janice Gould, eds. *Speak to Me Words: Essays on Contemporary American Indian Poetry*. Tucson: University of Arizona Press, 2003.

Rees, Martin. *Just Six Numbers: The Deep Forces That Shape the Universe*. New York: Basic Books, 2000.

Ro, Young-chan. *The Korean Neo-Confucianism of Yi Yulgok*. Albany: State University of New York Press, 1989.

Rosenberg, Tina. *The Haunted Land: Facing Europe's Ghosts after Communism*. New York: Vintage Books, 1996.

Schopenhauer, Arthur. *Essays and Aphorisms*. Translated by R. J. Hollingdale. New York: Penguin Classics, 1976.

Schwartz, Peter, and Doug Randall. *An Abrupt Climate Change Scenario and Its Implications for United States National Security*. Report prepared by the Global Business Network for the Department of Defense, October 2003. Available from http://www.gbn.com /ArticleDisplayServlet.srv?aid =26231.

Schweitzer, Albert. *Out of My Life and Thought: An Autobiography*. Translated by C. T. Campion. New York: New American Library, 1953.

Seneca, Lucius Annaeus. *Letters from a Stoic*. Translated by Robin Campbell. New York: Penguin Books, 1969.

Sherrard, Philip. *The Eclipse of Man and Nature: An Enquiry into the Origins and Consequences of Modern Science*. West Stockbridge, Mass.: Lindisfarne Press, 1987.

Snyder, Gary. *Axe Handles*. San Francisco: North Point Press, 1983.

———. *The Gary Snyder Reader: Prose, Poetry, and Translations, 1952–1998*. Washington, D.C.: Counterpoint, 1999.

———. *Turtle Island*. New York: New Directions, 1974.

A Source Book in Chinese Philosophy. Compiled and translated by Wing-tsit Chan. Princeton, N.J.: Princeton University Press, 1963.

St. John de Crèvecoeur, J. Hector. *Letters from an American Farmer and Sketches of Eighteenth-Century America*. Edited by Albert E. Stone. New York: Penguin Books, 1981.

Stafford, William. *The Way It Is: New and Selected Poems*. St. Paul, Minn.: Graywolf Press, 1999.

Teasdale, Sara. *The Collected Poems of Sara Teasdale*. New York: Macmillan, 1938.

Tucker, Mary Evelyn. *Moral and Spiritual Cultivation in Japanese Neo-Confucianism: The Life and Thought of Kaibara Ekken*. Albany: State University of New York Press, 1989.

Tucker, Mary Evelyn, and John Berthrong, eds. *Confucianism and Ecology: The Interrelation of Heaven, Earth, and Humans*. Cambridge, Mass.: Harvard University Center for the Study of World Religions, 1998.

Underhill, Evelyn. *The Spiritual Life*. New York: Harper, 1936.

The Upanishads. Translated by Juan Mascaró. New York: Penguin Books, 1979.

U.S. Department of the Army. *American Military History, 1607–1953*. Washington, D.C., 1956.

Wordsworth, William. *The Complete Poetical Works of William Wordsworth*. Edited by Andrew J. George. Boston: Houghton Mifflin, 1932.

World Commission on Environment and Development. *Our Common Future*. Oxford: Oxford University Press, 1987.

Yutang, Lin. *The Wisdom of Confucius*. New York: Modern Library, 1994.

Index